Mathematical Problem-Solving

Workbook 2

Strategies for Solving Real-World Problems

Satya Pradhan

Copyright © 2018 Satya Pradhan

All rights reserved. All rights reserved. No part of this publication may be reproduced or transmitted in any form or by any means, electronic or mechanical, including photocopy, or any information storage and retrieval system without permission from the copyright holder.

ISBN: 1541377559

ISBN-13: 978-1541377554

Contents

Acknowledgments v
Introduction vi
Assessment 1
1. Mathematical Operations 5
   1.1 Addition and Subtraction Key Words – 1 (*) 5
   1.2 Addition and Subtraction Key Words – 2 (*) 7
   1.3 Write a Math Expression (*) 9
   1.4 Write a Math Expression with Multiple Operations (*) 11
2. Basic Problem-Solving Strategies 15
   2.1 One-Step Problems – 1 (*) 15
   2.2 One-Step Problems – 2 (*) 17
   2.3 Multistep Problems (**) 19
   2.4 Work Backward – 1 (**) 21
   2.5 Work Backward – 2 (**) 23
   2.6 Too Much and Too Little Information – 1 (*) 25
   2.7 Too Much and Too Little Information – 2 (*) 27
   2.8 Review of Chapter 2 – 1 (**) 29
   2.9 Review of Chapter 2 – 2 (**) 31
3. Number Problems 33
   3.1 Place-Value Concepts – 1 (*) 33
   3.2 Place-Value Concepts – 2 (**) 35
   3.3 Different forms to write number (*) 37
   3.4 Review of Chapter 3 – 1 (***) 39
   3.5 Review of Chapter 3 – 2 (***) 41
4. Age Problems 43
   4.1 Age Problems in the Present (*) 43
   4.2 Age Problems in the Future (*) 45
   4.3 Age Problems in the Past (*) 47
   4.4 Review of Chapter 4 – 1 (*) 49
   4.5 Review of Chapter 4 – 2 (*) 51
5. Travel Problems 53
   5.1 Measuring Time (*) 53
   5.2 Problems on Elapsed Time – 1 (*) 55
   5.3 Problems on Elapsed Time – 2 (**) 57
   5.4 Travel Problems – 1 (*) 59
   5.5 Travel Problems – 2 (**) 61
   5.7 Review of Chapter 5 – 2 (**) 65
6. Money Problems 67
   6.1 Shopping Problems (*) 67
   6.2 Expense Planning (**) 69
   6.3 Investment Problems (***) 71

  6.4 Pricing Problems (***) 73
  6.5 Profit and Loss (***) 75
  6.6 Review of Chapter 6 (**) 77
7. Mixture Problems **79**
  7.1 Mixture Problems with Objects (*) 79
  7.2 Mixture Problem with solutions (*) 81
  7.3 Review of Mixture Problems (**) 83
  7.4 Filling or Emptying a Tank – 1 (*) 85
  7.5 Filling or Emptying a Tank – 2 (*) 87
8. Patterns **89**
  8.1 Basics of Number Patterns (*) 89
  8.2 Number Patterns in a Sequence (*) 91
  8.3 Continuing Patterns Using Letters (*) 93
  8.4 Find a Pattern (*) 95
  8.5 Review of Chapter 8 – 1 (*) 97
  8.6 Review of Chapter 8 – 2 (*) 99
Quiz **101**

Acknowledgments

My sincere appreciation and thanks to the following people for their feedback and suggestions while teaching these lessons as part of an after-school math-enrichment program: Susie Bierman, Reynaldo Lorenzana, Shalini Kinger, Arun Sahoo, and Alicia Lopez. I would also like to acknowledge the help of Soumya Sahoo, Kallala Bibhasa Giri, and Kamalesh Parhi during the preparation of this workbook. My sincere appreciation to my wife, Nishi, for her support while I was working on this book along with my hectic, full-time job in Silicon Valley, California. Special thanks to my son, Sougat, and daughter, Sarika, for their valuable feedback on different lessons. They were my first reviewers, connecting the lessons to their classroom in school.

Introduction

Having strong problem-solving skills can make a huge difference in one's career in the modern knowledge-based economy. Problems are at the center of what we do at work every day, whether one is developing a vaccine for the winter flu, creating an antivirus for the Internet, delivering lifesaving drugs to remote villages, maximizing profits for a company, or understanding the complex structure of our universe. This means that being an effective and confident problem solver is really important to one's success. Much of that confidence comes from having a good understanding of strategy and the tools to use when approaching a problem. Therefore, it is essential for students to develop the skills and techniques for problem-solving from an early age when they are in elementary school.

Conceptual understanding, procedural and computational skills, and application of concepts to real-life problems are three pillars of mathematics education. Conceptual understanding involves knowing what to do, procedural fluency requires knowing how to do it, and problem-solving focuses on solving a wide variety of complex, real-life problems using mathematical knowledge. Mathematical skills have been taught in school by placing maximum emphasis on the understanding of math concepts and computational skills, followed by applying the concepts to real-world problems. However, the real-life problem-solving requires students to apply these concepts in the exact opposite order, starting with understanding the problem, then finding the mathematical concepts required to solve it, and finally choosing the method that best solves the problem.

Mathematical problem-solving is often taught as a way to reinforce mathematical concepts, which misses the importance of strategic thinking while solving a problem. Many research articles and books have been written emphasizing the importance of problem-solving strategies. However, the burden of teaching problem-solving strategies is left mostly to teachers and parents, who are expected to develop their own curriculum and lesson plan for the complex topic of strategy and then teach it to students.

This book presents several problem-solving strategies that teachers and parents can easily use to teach the subject. The first two chapters present the concepts of number operations and the basic problem-solving strategies listed below:
- Solving one-step problems
- Solving multistep problems
- Solving problems working backward
- Formulating problems with variables and equations
- Solving problems using variables

Then the concept of the unitary method is presented in chapter 3. The remaining chapters present lessons about different problem types. The objective is to teach students how to start with a problem statement, understand the problem, and then solve it with a known mathematical procedure.

Students will encounter many different problem types in their careers. We have selected the following problem types that are appropriate for students in second grade:
- Number problems
- Age problems
- Travel problems
- Money problems
- Mixture problems
- Patterns

Each lesson in the workbook is classified as (*), (**), or (***), depending on the level of difficulty, and each starts with a few examples showing how to solve a particular type of problem. These are followed by a number of problems of this type. Students are expected to know the following basic computational skills in order to solve the problems in this workbook:
- Place-value concepts for decimals and whole numbers
- Operations with whole numbers, decimals, and fractions
- Proportion, ratio, and percentage
- Greatest common factor (GCF) and least common multiple (LCM)

Notes to Parents, Teachers, and Tutors

As a parent, you can use this workbook to teach problem-solving techniques to your child without any teaching experience. The first three chapters present the basic concepts and should be taught first. If students are already familiar with these concepts, these chapters can be skipped. All other chapters are independent of one another and can be taught in any order.

As a schoolteacher, you can easily integrate this workbook into the school curriculum by choosing appropriate lessons to be taught along with the curriculum.

Private tutors and after-school learning centers can use this workbook to offer special classes on mathematical problem-solving or as part of other math-enrichment programs. We suggest teaching two or three lessons a week, using the example questions provided at the beginning of each lesson and giving the other questions as homework.

Conceptual understanding of mathematical problem-solving is the main focus of this book. Therefore, we encourage students to use a calculator to solve the numerical expressions. This will allow them to take less time for numerical calculations and better focus on understanding concepts.

Answer Keys

Answer keys for all questions in this book are available online. You can download the PDF file for the answer keys at www.ilecy.com/BookAnswers

Feedback

We are always looking for feedback from students, parents, and teachers to make this book better. Please send your comments, testimonials, or suggestions for improvement to mathPS100@gmail.com.

Name _____   ▶ **Assessment**

Assessment

<u>Note</u>: Some of the questions in the assessment may be challenging for second-grade students.

Chapter 1:

1. What operation will you use for the key words **all together**?
   (a) Addition
   (b) Subtraction
   (c) None of the above

   Answer: _____

2. What is the math sentence for the following expression?

   Difference between 15 and 5 added to 25
   (a) 25 – (15 + 5)
   (b) 15 + (25 – 5)
   (c) (15 – 5) + 25
   (d) 25 + (15 + 5)

   Answer: _____

3. What is the operation key word in the following sentence?

   Luke's salary increased by $25.00.

   Answer: _____

Chapter 2:

4. A plant was 8 inches tall in March. It had grown by 7 inches every month from March through June. How tall is the plant in June?

   Answer: ____ _____
                 unit

5. Review the question given below and choose the best answer about the available information.

   Simon has 32 candies. He bought 10 more candies, 2 cakes, and 1 ice cream cone. How many candies does he have in total?

   (a) Too much information
   (b) Too little information
   (c) The right amount of information

   Answer: _____

6. Nikita spent $15.00 on a dress, $7.00 on a pair of sandals, and $5.00 on some cosmetics. After paying for all the items, she was left with $3.00. How much money did she have at the beginning?

   Answer: _____

Chapter 3:

7. I am given a number that can be written in expanded form, as given below:

   600 + 40 + 5

   If I change 600 to 800, what will be the new number in standard form?

   Answer: _____

1

Name _____   ▶ **Assessment**

8. In a 2-digit number, the tens digit is 8, and the ones digit is 2 less than the tens digit. What is the number?

    Answer: _____

9. Write 542 in expanded form and find the missing number in the following math sentence:

    542 = ____ + 40 + 2

    Answer: _____

10. What is the sum of the place values of 2 and 6 in 246?

    Answer: _____

11. What is the greatest possible 3-digit number using the digits 1, 9, and 3?

    Answer: _____

Chapter 4:

12. Jimmy is currently 7 years old. Kelvin is 5 years older than Jimmy. How old is Kelvin?

    Answer: _____

13. Sonia will be 17 years old in 2 years. How old is she now?

    Answer: _____

14. The sum of Ayush's and Brian's ages is 18. If Ayush is 8 years old, how old is Brian?

    Answer: _____

15. Pamela's current age is 19. If Juhi is 6 years younger than Pamela, how old is Juhi?

    Answer: _____

16. Craig will be 10 years old in 4 years. How old is he now?

    Answer: _____

17. Disha is 14 years old, and Anni is 12 years old. What is the sum of their current ages?

    Answer: _____

Chapter 5:

18. A song competition started at 8:00 a.m. and continued for 3 hours. When did the competition end?

    (a) 10:00 a.m.
    (b) 11:00 a.m.
    (c) 12:00 p.m.
    (d) 5:00 a.m.

    Answer: _____

Name _____ ▶ **Assessment**

19. If yesterday was Sunday, which day will it be tomorrow?

    Answer: _____

20. Mr. Turner went on a tour for 5 days. If his tour started on May 10, on which date will he come back?
    (a) May 15
    (b) May 18
    (c) May 12
    (d) May 17

    Answer: _____

21. George can walk a certain distance in 20 minutes. It takes Michael 10 minutes longer than George to walk the same distance. How long will Michael take to walk the same distance?

    Answer: ____ _____
    unit

22. It is 9 o'clock, and Rita is going to school. Is it nighttime or daytime?
    (a) Nighttime
    (b) Daytime

    Answer: _____

23. Patricia drove 60 kilometers in 2 hours. Anuj drove 80 kilometers in 2 hours. How much farther did Anuj drive than Patricia?

    Answer: ____ _____
    unit

Chapter 6:

24. Angela bought 6 novels for $40.00. After a few days, she wanted to sell those novels. She sold them for $35.00. How much of a loss did she take?

    Answer: _____

25. A pizza costs $7.00, a burger costs $4.00, and a pack of ice cream cones costs $10.00. How much money do we need to buy these three things?

    Answer: _____

26. The cost of 3 dresses is $82.00. Mia gave $90.00 to the cashier. How much money will the cashier return to her?

    Answer: _____

27. David deposited an amount of $60.00 in his savings account. After one year, he earned $12.00 in interest. What is the total amount in his account?

    Answer: _____

28. Frank bought a bicycle for $85.00 and sold it for $92.00. How much of a profit did he make?

    Answer: _____

Name _____ **Assessment**

Chapter 7:

29. Bottle 1 has 500 milliliters of water and 200 milliliters of alcohol. Bottle 2 has 250 milliliters of water and 100 milliliters of alcohol. If we mix the contents of both of the bottles, what is the total amount of alcohol?

    Answer: ____ _____
    unit

30. Pipe A can empty a water tank in 44 minutes. It takes Pipe B 12 minutes less than Pipe A to empty the same tank. How long does it take Pipe B to empty the water tank?

    Answer: ____ _____
    unit

31. Box 1 has 25 red marbles and 16 blue marbles. Box 2 has 32 blue marbles. If we mix the marbles from both of the boxes, what will be the number of blue marbles?

    Answer: ____ _____
    unit

32. Two nozzles are used for filling an oil drum. Nozzle 1 can add 15 liters of oil in a minute. Nozzle 2 can add 6 liters more than Nozzle 1 in a minute. How much oil can Nozzle 2 add in a minute?

    Answer: ____ _____
    unit

Chapter 8:

33. What is the next item in the following pattern?

    B1   C2   D3   ____

    Answer: _____

34. Alice wrote a number pattern as shown below:

    2, 4, 6, 8, ____

    What will be the next number in this pattern?

    Answer: _____

35. What is the next item in the following pattern?

    A   AB   ABC   ____

    Answer: _____

36. Look at the pattern: 1, 4, 7, 11, 13.

    There is one number that is wrong in the pattern.

    (a) The rule in this pattern is "Add ____ to each number."

    (b) The wrong number in the pattern = _____

    (c) The correct number should be = _____

    Answer: (a) ____ (b) ____ (c) ____

4

Name _____   ▶ Lesson 1.1

1. Mathematical Operations

1.1 Addition and Subtraction Key Words – 1 (*)

Example 1:

What is the addition or subtraction key word(s) in the following expression?

   Total of 20 and 35

Solution:

In this expression, **total of** are the addition key words.

Example 2:

What operation will you use for the key word **minus**?
  (a) Addition
  (b) Subtraction
  (c) None of the above

Solution:

The key word **minus** is a **subtraction** key word. So the answer is (b).

Write or choose the letter of the answer.

1. What is the operation key word(s) in the following sentence?

   Ram's monthly salary increased by $38.00.

   Answer: _____

2. What is the operation key word(s) in the following sentence?

   Rahul had 5 more crayons than her sister.

   Answer: _____

3. What operation will you use for the key word(s) in question 2?
   (a) Subtraction
   (b) Addition
   (c) Both (a) and (b)
   (d) None of the above

   Answer: _____

4. What operation will you use for the key words **take away**?
   (a) Subtraction
   (b) Addition
   (c) None of the above

   Answer: _____

5. What operation will you use for the key words **in all**?
   (a) Addition
   (b) Subtraction
   (c) Both (a) and (b)
   (d) None of the above

   Answer: _____

6. What is the operation key word(s) in the following expression?

   12 taken away from 65

   Answer: _____

Name _____ Lesson 1.1

Write or choose the letter of the answer.

7. What operation will you use for the key word(s) in question 6?
   (a) Addition
   (b) Subtraction
   (c) Both (a) and (b)
   (d) None of the above

   Answer: _____

8. What operation will you use for the key word **gain**?
   (a) Subtraction
   (b) Addition
   (c) Both (a) and (b)
   (d) None of the above

   Answer: _____

9. What is the operation key word(s) in the following sentence?

   Mr. Harper had a loss of $25.00 in the last month.

   Answer: _____

10. What operation will you use for the key word(s) in question 9?

    Answer: _____

11. What operation will you use for the key word **raise**?
    (a) Addition
    (b) Subtraction
    (c) None of the above

    Answer: _____

12. What is the operation key word(s) in the following sentence?

    How much money do Bill and Mark have altogether?

    Answer: _____

13. What operation will you use for the key word **remainder**?
    (a) Addition
    (b) Subtraction
    (c) Both (a) and (b)
    (d) None of the above

    Answer: _____

14. What is the operation key word(s) in the following expression?

    Difference between 85 and 72

    Answer: _____

15. What operation will you use for the key word(s) in question 14?

    Answer: _____

16. What operation will you use for the key words **total of**?
    (a) Subtraction
    (b) Addition
    (c) Both (a) and (b)
    (d) None of the above

    Answer: _____

Name _____   ▶ Lesson 1.2

1.2 Addition and Subtraction Key Words – 2 (*)

Example 1:

What operation will you use for the key word **loss**?
  (a) Subtraction
  (b) Addition
  (c) None of the above

Solution:

Loss is **subtraction** key word. So the answer is (a).

Example 2:

What is the addition or subtraction key word(s) in the following expression?

Sum of 150 and 92

Solution:

In this expression, **sum of** are the addition key words.

Write or choose the letter of the answer.

1. What operation will you use for the key word **less**?
   (a) Subtraction
   (b) Addition
   (c) None of the above

   Answer: _____

2. What is the operation key word(s) in the following sentence?

   How many apples and oranges are there in all?

   Answer: _____

3. What operation will you use for the key word(s) in question 2?
   (a) Subtraction
   (b) Addition
   (c) None of the above

   Answer: _____

4. What operation will you use for the key words **difference between**?
   (a) Subtraction
   (b) Addition
   (c) None of the above

   Answer: _____

5. What operation will you use for the key word **more**?
   (a) Addition
   (b) Subtraction
   (c) None of the above

   Answer: _____

6. What is the operation key word(s) in the following sentence?

   Nikki had a gain of $75.00 last year.

   Answer: _____

Name _____ ▶ Lesson 1.2

Write or choose the letter of the answer.

7. What operation will you use for the key words **amount of**?
   (a) Subtraction
   (b) Addition
   (c) None of the above

   Answer: _____

8. What is the operation key word(s) in the following expression?

   Sum of 32 and 9

   Answer: _____

9. What operation will you use for the key words **decreased by**?
   (a) Addition
   (b) Subtraction
   (c) None of the above

   Answer: _____

10. What is the operation key word(s) in the following expression?

    11 less than 25

    Answer: _____

11. What operation will you use for the key word(s) in question 10?
    (a) Subtraction
    (b) Addition
    (c) None of the above

    Answer: _____

12. What operation will you use for the key words **increase by**?
    (a) Addition
    (b) Subtraction
    (c) None of the above

    Answer: _____

13. What is the operation key word(s) in the following sentence?

    Rita had fewer candies than Sima.

    Answer: _____

14. What operation will you use for the key word **minus**?
    (a) Addition
    (b) Subtraction
    (c) None of the above

    Answer: _____

15. What is the operation key word(s) in the following sentence?

    The sum of 8 and 20 is 28.

    Answer: _____

16. What operation will you use for the key word **much**?
    (a) Subtraction
    (b) Addition
    (c) None of the above

    Answer: _____

Name _____   ▶ **Lesson 1.3**

1.3 Write a Math Expression (*)

Example 1:

What is the math sentence for the following problem?

Bob has 25 candies. Rob has 7 more candies than Bob. How many candies does Rob have?

(a) 25 − 7
(b) 25 + 7
(c) 7 − 25
(d) None of the above

Solution:

The following information is given:

Number of candies Bob has = 25

We can answer the question using the following math sentence:

(number of candies Rob has)
= (number of candies Bob has) + 7
= 25 + 7

So the answer is (b).

Example 2:

What is the operation key word(s) in the following expression?

18 more than 46

Solution:

In this expression, **more** is the addition key word.

Write or choose the letter of the answer.

1. What is the operation key word(s) in the following sentence?

    Angela has fewer pencils than Mia.

    Answer: _____

2. What is the operation key word(s) in the following sentence?

    Manoj had a gain of $200.00 in his annual income.

    Answer: _____

3. What is the math sentence for the following problem?

    Luke spent $22.00 for shirts and $50.00 for a backpack. How much money did he spend in total?

    (a) 22 + 50
    (b) 22 − 50
    (c) 50 + 20
    (d) None of the above

    Answer: _____

Name _____   Lesson 1.3

Write or choose the letter of the answer.

4. One pizza costs $7.00. Which math sentence will you use to find the cost of 2 pizzas?

   (a) 7 + 7
   (b) 7 × 7
   (c) 7 – 7
   (d) All of the above

   Answer: _____

5. What is the math sentence for the following problem?

   Ana bought 15 candies. She ate 8 of them and gave the remaining candies to her sister. How many candies did she give to her sister?

   (a) 15 × 10
   (b) 15 + 8
   (c) 15 – 8
   (d) All of the above

   Answer: _____

6. What operation will you use for the key words **increase of**?

   (a) Addition
   (b) Subtraction
   (c) None of the above

   Answer: _____

7. What is the operation key word(s) in the following expression?

   75 taken away from 90

   Answer: _____

8. What is the operation key word(s) in the following sentence?

   Emily has more toys than her sister.

   (a) Than
   (b) More
   (c) None of the above

   Answer: _____

9. What is the operation key word(s) in the following question?

   What is the difference between 50 and 35?

   Answer: _____

10. What is the math sentence for question 9?

    (a) 50 – 45
    (b) 35 ÷ 50
    (c) 50 + 35
    (d) 50 – 35

    Answer: _____

11. Amit had 10 marbles. He bought 25 more marbles. Which math sentence will you use to find the total number of marbles that Amit has?

    (a) 10 + 25
    (b) 10 – 25
    (c) 25 – 10
    (d) All of the above

    Answer: _____

Name _____ Lesson 1.4

1.4 Write a Math Expression with Multiple Operations (*)

Example 1:

What is the math sentence for the following expression?

10 plus 15 minus 5

(a) (10 + 15) − 5
(b) (10 + 5) − 15
(c) (15 − 10) + 5
(d) (15 − 5) + 10

Solution:

We can write the math sentence as follows:

- 10 plus 15 = 10 + 15
- (10 plus 15) minus 5
   = (10 + 15) minus 5
   = (10 + 15) − 5

So the answer is (a).

Example 2:

What is the math sentence for the following expression?

8 subtracted from the sum of 20 and 35

(a) (20 − 8) − 35
(b) (35 − 20) + 8
(c) (20 + 35) − 8
(d) (20 − 8) + 35

Solution:

We can write the math sentence as follows:

- The sum of 20 and 35 = 20 + 35
- 8 subtracted from the sum of 20 and 35 = (the sum of 20 and 35) minus 8
   = (20 + 35) − 8

So the answer is (c).

Choose the letter of the answer.

1. What is the math sentence for the following expression?

   5 more than the sum of 15 and 7

   (a) (15 − 7) + 5
   (b) (15 + 7) + 5
   (c) (15 + 7) − 5
   (d) (5 − 7) + 15

   Answer: _____

2. What is the math sentence for the following expression?

   Sum of 10 and 22 is subtracted from 55

   (a) 55 − (10 + 22)
   (b) 55 + (10 + 22)
   (c) 55 − (10 − 22)
   (d) 55 + (22 − 10)

   Answer: _____

Name _____

Lesson 1.4

Choose the letter of the answer.

3. What is the math sentence for the following expression?

   9 less than the difference between 14 and 3

   (a) (14 − 3) − 9
   (b) (14 − 9) − 3
   (c) (14 + 9) − 3
   (d) (14 + 3) − 9

   Answer: _____

4. What is the math sentence for the following expression?

   Total of 60 and 45 minus 5

   (a) (45 + 5) − 60
   (b) (60 − 5) + 45
   (c) (60 + 45) + 5
   (d) (60 + 45) − 5

   Answer: _____

5. What is the math sentence for the following expression?

   8 more than the difference between 10 and 6

   (a) (10 + 8) − 6
   (b) (10 − 6) + 8
   (c) (8 + 6) − 10
   (d) (10 + 6) − 8

   Answer: _____

6. What is the math sentence for the following expression?

   4 taken away from the total of 20 and 32

   (a) (32 + 4) − 20
   (b) (20 + 32) − 4
   (c) (20 + 4) + 32
   (d) 32 − (4 + 20)

   Answer: _____

7. What is the math sentence for the following expression?

   Sum of 42 and 64 decreased by 10

   (a) (42 + 64) − 10
   (b) (64 + 42) + 10
   (c) (64 + 10) − 42
   (d) 42 + (64 − 10)

   Answer: _____

8. What is the math sentence for the following expression?

   Total of 25 and 18 taken away from 60

   (a) 60 + (25 + 18)
   (b) (25 − 18) − 60
   (c) 60 − (25 + 18)
   (d) (60 +18) + 25

   Answer: _____

Name _____   Lesson 1.5

1.5 Review of Chapter 1 (*)

Write or choose the letter of the answer.

1. What operation will you use for the key words **decreased by**?
   (a) Subtraction
   (b) Addition
   (c) None of the above

   Answer: _____

2. What is the operation key word(s) in the following expression?

   12 plus 36

   Answer: _____

3. What is the operation key word(s) in the following sentence?

   Pamela has 3 fewer pencils than Juhi.

   Answer: _____

4. What is the math sentence for the following problem?

   Elina made 50 paper crafts. She gave 8 paper crafts to her friend Ria. How many paper crafts did she have after giving 8 to Ria?

   (a) 8 × 50
   (b) 50 + 8
   (c) 50 − 8
   (d) All of the above

   Answer: _____

5. What is the operation key word(s) in the following question?

   What is the sum of 100 and 75?

   Answer: _____

6. What operation will you use for the key word **remainder**?
   (a) Addition
   (b) Subtraction
   (c) None of the above

   Answer: _____

7. What is the math sentence for the following expression?

   Total of 28 and 34 subtracted from 70

   (a) 70 − (28 + 34)
   (b) 70 + (34 − 28)
   (c) (28 + 34) − 70
   (d) 70 − (34 − 28)

   Answer: _____

8. Raul had $24.00. His father gave him $10.00. Which math sentence will you use to find the total amount of money Raul has?
   (a) $24.00 − $10.00
   (b) $10.00 − $24.00
   (c) $24.00 + $10.00
   (d) All of the above

   Answer: _____

Name _____   ▶ Lesson 1.5

Write or choose the letter of the answer.

9. What is the operation key word(s) in the following sentence?

   Nikhil's age is increased by 5 in 5 years.

   Answer: _____

10. What operation will you use for the key word(s) in question 9?
   (a) Addition
   (b) Subtraction
   (c) Both a and b
   (d) None of the above

   Answer: _____

11. What is the math sentence for the following expression?

   45 less than the sum of 55 and 82
   (a) (45 + 55) − 82
   (b) (82 − 55) + 45
   (c) (55 + 82) − 45
   (d) (45 + 82) − 55

   Answer: _____

12. What operation will you use for the key word **gain**?
   (a) Subtraction
   (b) Addition
   (c) Both (a) and (b)
   (d) None of the above

   Answer: _____

13. What is the math sentence for the following problem?

   Julie bought 35 crayons to make a drawing. She used 22 of them, and the rest of them she kept for future use. How many crayons did she keep for later?
   (a) 35 + 22
   (b) 35 + 25
   (c) 35 − 22
   (d) None of the above

   Answer: _____

14. What is the operation key word(s) in the following sentence?

   Jacob had a remainder of $12.00 after returning from a picnic.

   Answer: _____

15. What operation will you use for the key word(s) in question 14?
   (a) Subtraction
   (b) Addition
   (c) None of the above

   Answer: _____

16. What operation will you use for the key word **altogether**?
   (a) Addition
   (b) Subtraction
   (c) None of the above

   Answer: _____

2. Basic Problem-Solving Strategies

2.1 One-Step Problems – 1 (*)

Example 1:

Andy has 27 marbles. Julie has 15 marbles. What is the total number of marbles that they have?

Solution:

The following information is given:

Number of marbles Andy has = 27

Number of marbles Julie has = 15

The total number of marbles they have

= (number of marbles Andy has)

+ (number of marbles Julie has)

= 27 + 15

= 42

So they have 42 marbles in total.

Example 2:

Rob had 22 crayons. He gave 8 crayons to his sister. How many crayons did Rob have left?

Solution:

The following information is given:

Number of crayons Rob had = 22

Number of crayons Rob gave to his sister = 8

The number of crayons Rob had left

= (number of crayons Rob had)

– (number of crayons he gave to his sister)

= 22 – 8

= 14

So Rob has 14 crayons left.

Write or choose the letter of the answer.

1. Kavya has 5 pens. She gave 3 pens to her sister. What operation will you use to find the number of pens Kavya has left?

    (a) Subtraction
    (b) Addition
    (c) Both (a) and (b)
    (d) None of the above

    Answer: _____

2. What is the operation key word(s) in the following problem?

    Sam has 26 toys. Juhi has 19 more toys than Sam. How many toys does Juhi have?

    (a) More
    (b) Toys
    (c) Many
    (d) None of the above

    Answer: _____

Name _____

Lesson 2.1

Write or choose the letter of the answer.

3. Allen has 16 mangoes. David has 9 fewer mangoes than Allen. How many mangoes does David have?

   Answer: ____ _____
                           unit

4. Raul has 25 chocolates. He gave 10 chocolates to Rohit. How many chocolates did Raul have left?

   Answer: ____ _____
                           unit

5. Mark has 16 candies. He gave 8 candies to his friend. What operation will you use to find the number of candies that Mark has left?

   (a) Subtraction
   (b) Addition
   (c) Both (a) and (b)
   (d) None of the above

   Answer: _____

6. Mia has 10 books on her shelf. She added 5 more books to her shelf. How many books does Mia now have?

   Answer: ____ _____
                           unit

7. Jacob has 22 flowers. He gave 16 flowers to a friend. What operation will you use to find the number of flowers Jacob has left?

   (a) Addition
   (b) Subtraction
   (c) Multiplication
   (d) None of the above

   Answer: _____

8. What is the operation key word(s) in the following problem?

   Nancy has 13 apples. Her brother gave her 3 more apples. How many apples does Nancy now have?

   (a) Subtraction
   (b) Addition
   (c) Both (a) and (b)
   (d) None of the above

   Answer: _____

9. Nelson has 40 sticks of chalk. He gave 20 sticks of chalk to his sister. How many sticks of chalk does Nelson now have?

   Answer: ____ _____
                           unit

Name _____   ▶ Lesson 2.2

2.2 One-Step Problems – 2 (*)

Example 1:

Andy spent $90.00 on a watch and $75.00 on a pair of sunglasses. How much money did he spend in total?

Solution:

The following information is given:
  Amount spent on watch = $90.00
  Amount spent on sunglasses = $75.00

Total amount spent
  = (amount spent on watch)
    + (amount spent on sunglasses)
  = $90.00 + $75.00
  = $165.00

So Andy spent $165.00 in total.

Example 2:

Jacob had $60.00. He gave $20.00 to his sister. How much money did Jacob have left?

Solution:

The following information is given:
  Amount of money Jacob had = $60.00
  Amount of money he gave to his sister = $20.00
Amount of money he had left
  = (amount of money Jacob had)
    − (amount of money he gave to his sister)
  = $60.00 − $20.00
  = $40.00

So Jacob had $40.00 left.

Write or choose the letter of the answer.

1. Rishab had $300.00 in his savings account. He earned $120.00 in interest after several years. What operation will you use to find the total amount of interest that he earned?

   (a) Subtraction
   (b) Addition
   (c) Both (a) and (b)
   (d) None of the above

   Answer: _____

2. What is the operation key word(s) in the following problem?

   Aryan had $95.00. Amit had $25.00 less than Aryan. How much money did Amit have?

   (a) Money
   (b) Had
   (c) Less
   (d) None of the above

   Answer: _____

Name _____   ▶ Lesson 2.2

Write or choose the letter of the answer.

3. Mr. Clark bought some colored paint. He paid $8.00 for red paint and $9.00 for green paint. How much money did he give to the cashier?

   Answer: _____

4. Elina spent $20.00 in a shop and $35.00 in a hotel. How much money did she spend in total?

   Answer: _____

5. Neha has to make 21 circles in a paper. She made 15 of them in the morning. What operation will you use to find how many more circles need to be made?

   (a) Subtraction
   (b) Addition
   (c) Both (a) and (b)
   (d) None of the above

   Answer: _____

6. Bill spent $65.00 on a textbook and $12.00 on a T-shirt. How much money did he spend in total?

   Answer: _____

7. Rohan had $17.00. His brother had $11.00 more than Rohan. How much money did his brother have?

   Answer: _____

8. What is the operation key word(s) in the following problem?

   Neha has 13 mangoes. She gave 4 mangoes to her sister. How many mangoes does Neha now have?

   (a) Subtraction
   (b) Addition
   (c) Both (a) and (b)
   (d) None of the above

   Answer: _____

9. Disha had $12.00 to buy flowers. She spent $7.00 on roses and the rest on jasmine. How much money did Disha spend on jasmine?

   Answer: _____

10. Rohit had $23.00. His sister had $15.00 less than Rohit. How much money did his sister have?

    Answer: _____

Name _____   ▶ Lesson 2.3

2.3 Multistep Problems (**)

Example 1:

Jack spent $12.00 on a burger, $15.00 on a pizza, and $21.00 on an ice cream cone. After paying for the food, he had $9.00 left. How much money did he have at the beginning?

Solution:

The following information is given:
    Money spent on a burger = $12.00
    Money spent on a pizza = $15.00
    Money spent on an ice cream cone = $21.00

The total amount of money spent
    = (money spent on a burger)
    + (money spent on a pizza)
    + (money spent on an ice cream cone)
    = $12.00 + $15.00 + $21.00
    = $48.00

Amount left after Jack paid for the food = $9.00

Amount of money at the beginning
    = (total amount of money spent)
        + (amount left)
    = $48.00 + $9.00
    = $57.00

So Jack had $57.00 at the beginning.

Example 2:

Amar spent $9.00 on chocolates and $8.00 on biscuits. If he gave $20.00 to the cashier, how much money did the cashier return?

Solution:

The following information is given:
    Money spent on chocolates = $9.00
    Money spent on biscuits = $8.00

The total amount of money spent
    = (money spent on chocolates)
    + (money spent on biscuits)
    = $9.00 + $8.00
    = $17.00

Money given to the cashier = $20.00

Amount of money the cashier returned
    = (money given to the cashier)
    − (total amount of money spent)
    = $20.00 − $17.00
    = $3.00

So the cashier returned $3.00.

Write the answer.

1. Mr. Lee bought some books. He paid $18.00 for a novel and $8.00 for a storybook. After paying for the books, he had $4.00 left. How much did he have at the beginning?

    Answer: _____

2. Nathan had $22.00. He spent $6.00 on white socks, $8.00 on brown socks, and the rest on black socks. How much money did Nathan spend on black socks?

    Answer: _____

Lesson 2.3

Write the answer.

3. Kunal had $20.00. He spent $8.00 on shoes, $7.00 on track pants, and the rest on deodorant. How much money did he spend on deodorant?

    Answer: _____

4. Kapil spent $10.00 on a belt and $15.00 on a wallet. If he gave $30.00 to the cashier, how much money did the cashier return?

    Answer: _____

5. Vinod had $35.00. He spent $20.00 on dress shirts and $10.00 on books. How much money did he have left?

    Answer: _____

6. Robin had $41.00. He spent $9.00 on shoes, $21.00 on a bag, and the rest at a coffee shop. How much money did he spend at the coffee shop?

    Answer: _____

7. Binay had $46.00. He spent $18.00 on a bat, $9.00 on a ball, and the rest on other equipment. How much money did he spend on the other equipment?

    Answer: _____

8. Daniel went to a restaurant. He spent $15.00 on food and $12.00 on soft drinks. If he gave $30.00 to the cashier, how much money did the cashier return?

    Answer: _____

9. Mrs. Maria spent $6.00 on deodorant and $10.00 on perfume. After paying for them, she was left with $8.00. How much did she have at the beginning?

    Answer: _____

10. Jacob spent $41.00 on a microwave and $60.00 on a refrigerator. If he gave $115.00 to the cashier, how much money did the cashier return?

    Answer: _____

2.4 Work Backward – 1 (**)

Example 1:

Kapil bought 3 shirts for $45.00. If the cashier returned $5.00, how much money did Kapil give to the cashier?

Solution:

The following information is given:

Cost of 3 shirts = $45.00
Amount the cashier returned = $5.00

We can find the amount Kapil gave to the cashier as follows:

Amount given to the cashier

= (cost of 3 shirts)
+ (amount the cashier returned)
= $45.00 + $5.00 = $50.00

So Kapil gave $50.00 to the cashier.

Example 2:

Workers cleaned 20 rooms in 2 days. If they cleaned 8 rooms the first day, how many did they clean the second day?

Solution:

The following information is given:

Number of rooms cleaned in 2 days = 20
Number of rooms cleaned the first day = 8

We can find the number of rooms cleaned the second day as follows:

Number of rooms cleaned the second day
= (number of rooms to clean)
− (number of rooms cleaned the first day)
= 20 − 8
= 12

So they cleaned 12 rooms the second day.

Example 3:

Sam had $20.00. He spent $9.00 on flowers, $8.00 on pens, and the rest on fruit. How much did he spend on fruit?

Solution:

The following information is given:

Amount of money Sam had = $20.00
Money he spent on flowers = $9.00
Money he spent on pens = $8.00

We can find the amount Sam spent as follows:

- Find the total amount of money he spent on flowers and pens.

 (money spent on flowers and pens)
 = $9.00 + $8.00
 = $17.00

- Find the money spent on fruit.

 (amount of money spent on fruit)

 = (total amount Sam had) − (money spent on flowers and pens)
 = $20.00 − $17.00
 = $3.00

So Sam spent $3.00 on fruit.

Name _____ Lesson 2.4

Write the answer.

1. A plant was 20 inches tall on Monday. It grew 4 inches every day. How tall was the plant on Sunday?

   Answer: ____ _____
                    unit

2. Ritika had $36.00. She spent $15.00 on a pair of sunglasses, $10.00 on a sling bag, and the rest on a pair of sandals. How much money did she spend on the sandals?

   Answer: _____

3. Jack bought 5 books for $34.00 from a store. If the cashier returned $6.00, how much money did he give to the cashier?

   Answer: _____

4. 2 tailors stitched 35 dresses in 2 months. If they stitched 18 dresses in the first month, how many dresses did they stitch in the second month?

   Answer: ____ _____
                    unit

5. Raul was 45 inches tall in 2015. He grows 4 inches every year. How tall was he in 2016?

   Answer: ____ _____
                    unit

6. Max had $30.00. He spent $20.00 on shoes, $6.00 on movie tickets, and the rest on a wallet. How much money did he spend on the wallet?

   Answer: _____

7. John was 52 inches tall in 2014. He grew 15 inches between 2010 and 2014. How tall was he in 2010?

   Answer: ____ _____
                    unit

8. Workers watered 85 plants in 2 hours. If they watered 50 plants in the first hour, how many plants did they water in the second hour?

   Answer: ____ _____
                    unit

9. Ria bought 8 gifts for $48.00 from a store. If the cashier returned $7.00, how much money did she give to the cashier?

   Answer: _____

10. Jacob had $15.00. He spent $5.00 on pizzas, $4.00 on burgers, and the rest on ice cream cones. How much money did he spend on ice cream cones?

    Answer: _____

Name _____   Lesson 2.5

2.5 Work Backward – 2 (**)

Example 1:

Jack bought 2 pairs of jeans for $58.00 from a store. If the cashier returned $2.00, how much money did he give to the cashier?

Solution:

The following information is given:

Cost of 2 pairs of jeans = $58.00
Amount the cashier returned = $2.00

We can find the amount Jack gave the cashier as follows:

Amount Jack gave the cashier

= (cost of 2 pairs of jeans)
+ (amount the cashier returned)
= $58.00 + $2.00 = $60.00

So Jack gave $60.00 to the cashier.

Example 2:

Lucy was 57 inches tall in 2016. She grew by 3 inches between 2012 and 2016. How tall was she in 2012?

Solution:

The following information is given:

Height in 2016 = 57 inches
Height grown between 2012 and 2016
= 3 inches

We can find Lucy's height in 2012 as follows:

Lucy's height in 2012

= (height in 2016) – (height grown between 2012 and 2016)

= 57 – 3 = 54 inches

So Lucy was 54 inches tall in 2012.

Write the answer.

1. Kamron had $45.00. He spent $30.00 on a mobile phone and the rest on a keyboard. How much money did he spend on the keyboard?

    Answer: _____

2. A plant was 10 inches tall in June. It grew 4 inches between April and June. How tall was the plant in April?

    Answer: ____ _____
    unit

3. Workers washed 30 cars in 3 days. If they washed 19 cars in the first 2 days, how many cars did they wash the third day?

    Answer: ____ _____
    unit

4. Nikhil had $22.00. He spent $12.00 on his friends, and the rest he gave to his sister. How much money did he give to his sister?

    Answer: _____

Name _____

Lesson 2.5

Write the answer.

5. Some students made 7 projects in 3 months. If they made 4 projects in the first two months, how many projects did they make in the third month?

   Answer: ____ _____
   unit

6. Anil had $25.00. He spent $8.00 on a belt and the rest on a remote-controlled car. How much money did he spend on the remote-controlled car?

   Answer: _____

7. Catherine bought 3 pairs of jeans for $47.00 from a store. If the cashier returned $3.00, how much money did she give to the cashier?

   Answer: _____

8. Peter was 32 inches tall in 2005. He grew by 5 inches between 2002 and 2005. How tall was he in 2002?

   Answer: ____ _____
   unit

9. Sofia bought 3 pairs of headsets for $62.00 from a store. If the cashier returned $8.00, how much money did she give to the cashier?

   Answer: _____

10. Two friends wrote 55 pages in 2 days. If they wrote 32 pages on the first day, how many pages did they write on the second day?

    Answer: ____ _____
    unit

11. A horse was 6 feet tall in 2013. It grew by 2 feet between 2009 and 2013. How tall was the horse in 2009?

    Answer: ____ _____
    unit

12. Jasmine had $42.00. She spent $20.00 on a dress and the rest on a sari. How much money did she spend on the sari?

    Answer: _____

Name _____   ▶ **Lesson 2.6**

2.6 Too Much and Too Little Information – 1 (*)

Example 1:

If the following question has enough information, find the answer. Otherwise, write "No answer" for the answer.

There are 98 cakes and 94 burgers in a shop. If 80 cakes are sold, how many cakes are left?

Solution:

The following information is given:

Number of cakes in the shop = 98
Number of burgers = 94
Number of cakes sold = 80

You can find the number of cakes that are left as follows:

Number of cakes left
= (number of cakes in the shop)
− (number of cakes sold)
= 98 − 80
= 18 cakes

So there are 18 cakes left in the shop.

Example 2:

Review the question given below and choose the best answer about the available information.

Jay wants to buy a computer. How much money does he need to pay for the computer?

(a) Too little information
(b) Too much information
(c) The right amount of information

Solution:

Jay wants to buy a computer, but the price of the computer is not given in the question.

Without knowing the cost of the computer, we cannot find the amount of money that is needed for the computer.

So the answer is too little information, which is option (a).

Write or choose the letter of the answer.

1. Review the question given below and choose the best answer about the available information.

   Nikhil has 6 notebooks. He bought 8 more notebooks, 5 candies, and 1 laptop. How many notebooks does he have in total?

   (a) Too little information
   (b) Too much information
   (c) The right amount of information

   Answer: _____

2. Review the question given below and choose the best answer about the available information.

   Alice had 6 pairs of jeans. She bought 2 more pairs of jeans from a store. How many pairs of jeans does she have in total?

   (a) Too little information
   (b) Too much information
   (c) The right amount of information

   Answer: _____

Name _____   ▶ **Lesson 2.6**

Write or choose the letter of the answer.

3. If the following question has enough information, find the answer. Otherwise, write "No answer" for the answer.

    Bimal bought 3 ice cream cones. How much money did he give to the cashier?

    Answer: _____

4. Review the question given below and choose the best answer about the available information.

    Mia made 15 toys. She gave 7 toys to her sisters. How many toys does she have left?

    (a) Too much information
    (b) Too little information
    (c) The right amount of information

    Answer: _____

5. Review the question given below and choose the best answer about the available information.

    Pamela had 25 candies. She bought 7 more candies and 5 packages of chips from a store. How many candies does she have in total?

    (a) Too little information
    (b) Too much information
    (c) The right amount of information

    Answer: _____

6. Review the question given below and choose the best answer about the available information.

    Simran has more crayons than her brother Anuj. How many crayons does Simran have?

    (a) Too little information
    (b) The right amount of information
    (c) Too much information

    Answer: _____

7. If the following question has enough information, find the answer. Otherwise, write "No answer" for the answer.

    Simon has 18 shirts. If his father bought 3 more shirts for him on his birthday, how many shirts does he have in total?

    Answer: _____

8. Review the question given below and choose the best answer about the available information.

    Sarah has 25 marbles. She bought 10 more marbles, 8 ribbons, and 1 pencil box. How many marbles does she have in total?

    (a) The right amount of information
    (b) Too little information
    (c) Too much information

    Answer: _____

Name _____   ▶ Lesson 2.7

2.7 Too Much and Too Little Information – 2 (*)

Example 1:

Review the question given below and choose the best answer about the available information.

Ava has 6 teddy bears. She bought 4 more teddy bears, 3 burgers, and 1 package of cookies. How many teddy bears does she have in total?

(a) Too little information
(b) Too much information
(c) The right amount of information

Solution:
The following information is given:

Number of teddy bears Ava has = 6
Number of teddy bears Ava bought = 4
Number of burgers = 3
Number of packages of cookies = 1

You can find the total teddy bears by adding the teddy bears Ava had and the number of teddy bears she bought.

The information about burgers and packages of cookies is not required to solve this problem.

So the answer is <u>too much information,</u> which is option (b).

Example 2:

Review the question given below and choose the best answer about the available information.

Kate has 4 toy cars. She bought 2 more toy cars. How many toy cars does she have in total?

(a) Too little information
(b) Too much information
(c) The right amount of information

Solution:
The following information is given:

Number of toy cars Kate has = 4
Number of toy cars Kate bought = 2

You can find the total toy cars by adding the toy cars Kate had and the number of toy cars she bought.

Total number of toy cars Kate has
= 4 + 2 = 6 toy cars

The information in the question is sufficient to find the answer.

So the answer is <u>the right amount of information,</u> which is option (c).

Write or choose the letter of the answer.

1. If the following question has enough information, find the answer. Otherwise, write "No answer" for the answer.

   Angela bought 5 comic books from a book store. How much money did she pay the cashier?

   Answer: _____

2. If the following question has enough information, find the answer. Otherwise, write "No answer" for the answer.

   There are 128 water bottles and 97 Coke bottles. If 80 Coke bottles are sold, how many Coke bottles are left?

   Answer: _____

Name _____   Lesson 2.7

Write or choose the letter of the answer.

3. Review the question given below and choose the best answer about the available information.

   Pamela wants to buy an air conditioner. How much money does she need to pay for the air conditioner?

   (a) Too little information
   (b) Too much information
   (c) The right amount of information

   Answer: _____

4. If the following question has enough information, find the answer. Otherwise, write "No answer" for the answer.

   70 employees are working at a company. If 30 employees went on leave, how many employees were left at the company?

   Answer: _____

5. Review the question given below and choose the best answer about the available information.

   Bob has 4 toys. His father bought him 2 more toys, 3 pens, and 1 shirt. How many toys does Bob have in total?

   (a) Too little information
   (b) Too much information
   (c) The right amount of information

   Answer: _____

6. If the following question has enough information, find the answer. Otherwise, write "No answer" for the answer.

   Mia ate more mangoes than her friend Sarah. How many mangoes did Mia eat?

   Answer: _____

7. Review the question given below and choose the best answer about the available information.

   David has 15 crayons. He gave 5 crayons to a friend. How many crayons does he have left?

   (a) The right amount of information
   (b) Too little information
   (c) Too much information

   Answer: _____

8. Review the question given below and choose the best answer about the available information.

   Christina wants to buy a microwave. How much money does she need to pay for the microwave?

   (a) Too much information
   (b) The right amount of information
   (c) Too little information

   Answer: _____

2.8 Review of Chapter 2 – 1 (**)

Write or choose the letter of the answer.

1. Ryan bought 2 wallets for $36.00 from a store. If the cashier returned $4.00, how much money did he give to the cashier?

    Answer: _____

2. Sofia was 63 inches tall in 2015. She grew 15 inches between 2008 and 2015. How tall was she in 2008?

    Answer: ____ _____
            unit

3. Bob plucked 32 flowers. He gave 18 flowers to his aunt. How many flowers did Bob have left?

    Answer: ____ _____
            unit

4. Joseph has 25 colored pencils. He bought 12 more colored pencils from a store. What operation will you use to find the total number of colored pencils Joseph has?
    (a) Multiplication
    (b) Addition
    (c) Subtraction
    (d) None of the above

    Answer: _____

5. Julie spent $7.00 on a pizza, $5.00 on popcorn, and $13.00 on juice in a theater. After paying, she was left with $5.00. How much money did she have at the beginning?

    Answer: _____

6. If the following question has enough information, find the answer. Otherwise, write "No answer" for the answer.

    There are 35 birds sitting in a tree. If 15 birds fly away, how many birds are left?

    Answer: _____

7. Maria spent $65.00 in total. She spent $22.00 on a sari, $12.00 on a pair of sandals, and the rest on a dinner set. How much money did she spend on the dinner set?

    Answer: _____

8. Nisha had $15.00. She spent $9.00 on a movie ticket and the rest on snacks. How much money did Nisha spend on snacks?

    Answer: _____

Name _____   ▶ Lesson 2.8

Write or choose the letter of the answer.

9. Adriana bought 4 jumpsuits for $68.00 from a store. If the cashier returned $7.00, how much money did she give to the cashier?

   Answer: _____

10. Review the question given below and choose the best answer about the available information.

    Alex has 15 books. His father bought 7 more books, 5 pens, and 1 backpack. How many books does Alex have in total?
    (a) Too little information
    (b) Too much information
    (c) The right amount of information

    Answer: _____

11. Basket 1 has 25 apples. Basket 2 has 11 more apples than Basket 1. What operation will you use to find the total number of apples in Basket 2?
    (a) Addition
    (b) Multiplication
    (c) Subtraction
    (d) None of the above

    Answer: _____

12. A tap filled 32 buckets in 2 hours. If it filled 16 buckets in the first hour, how many buckets did it fill in the second hour?

    Answer: _____ _____
                        unit

13. George had 24 pencils. He gave 8 pencils to his brother, 10 pencils to his sister, and kept the rest.

    If you want to find the number of pencils George kept, what question do you need to answer first?
    (a) How many pencils did George have at first?
    (b) How many total pencils did he give to his brother and sister?
    (c) How many pencils does he have left?
    (d) All of the above

    Answer: _____

14. How many pencils does George have left in question 13?

    Answer: _____ _____
                        unit

15. A plant was 10 inches tall on Sunday. It grew 2 inches every day. How tall was the plant on Saturday?

    Answer: _____ _____
                        unit

16. Kevin bought 3 backpacks for $43.00 from a store. If the cashier returned $7.00, how much money did he give to the cashier?

    Answer: _____

2.9 Review of Chapter 2 – 2 (**)

Write or choose the letter of the answer.

1. Angela bought 3 headphones for $42.00 from a store. If the cashier returned $8.00, how much money did she give to the cashier?

    Answer: _____

2. Review the question given below and choose the best choice for the available information.

    Mia has 25 candies. She bought 15 more candies and 3 packages of cookies. How many candies does she have in total?
    (a) too little information
    (b) too much information
    (c) the right amount of information

    Answer: _____

3. Rakesh is 5 years younger than Anil. Anil is 22 years old now. What operation will you use to find Rakesh's current age?
    (a) Addition
    (b) Subtraction
    (c) Multiplication
    (d) None of the above

    Answer: _____

4. Peter bought 12 pizzas from a store. He gave 3 pizzas to his friend. How many pizzas were left with Peter?

    Answer: ____ _____

Name _____   Lesson 2.9

Write or choose the letter of the answer.

9. Sarah had 35 candies. She ate 7 of them and gave 15 candies to her brother.

   If you want to find the number of candies that Sarah kept, what question do you need to answer first?

   (a) How many candies does Sarah have left?
   (b) How many candies did Sarah have at first?
   (c) How many total candies did Sarah eat and give to her brother?
   (d) All of the above

   Answer: _____

10. How many candies did Sarah have left in question 9?

    Answer: _____ _____
                      unit

11. A plant was 15 inches tall in May. It grew 4 inches every month. How tall was the plant in March?

    Answer: _____ _____
                      unit

12. Jacob bought 5 containers of deodorant for $12.00 from a store. If the cashier returned $3.00, how much money did he give to the cashier?

    Answer: _____

13. Eric bought 2 jackets for $58.00 from a store. If the cashier returned $2.00, how much money did he give to the cashier?

    Answer: _____

14. If the following question has enough information, find the answer. Otherwise, write "No answer" for the answer.

    Nancy bought 5 bottles of juice from a store. How much money did she give to the cashier?

    Answer: _____

15. Vimal has 17 balloons. He bought 20 more balloons from a store. What operation will you use to find the total number of balloons Vimal has?

    (a) Addition
    (b) Multiplication
    (c) Subtraction
    (d) None of the above

    Answer: _____

16. Sonia had $32.00. She spent $13.00 on a set of cups and the rest on groceries. How much money did Sonia spend on groceries?

    Answer: _____

3. Number Problems

3.1 Place-Value Concepts – 1 (*)

Example 1:

    I am the largest 2-digit number with 6 as my ones digit. What number am I?

Solution:

We can find the answer by using a place-value table.

Tens	Ones

- The number has two digits. So there should be two boxes.

- The number is the largest 2-digit number, and 6 is the ones digit. This means we have 6 in the ones place. So write 6 in the ones place.

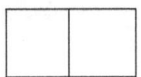

- The tens place has to be filled with the largest 1-digit number, which is 9.

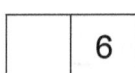

So the number is 96.

Example 2:

    What is the sum of the **values** of 3 and 8 in 358?

Solution:

In the number 358:

    Value of 3 = 300
    Value of 8 = 8

Sum of the values of 3 and 8 = 300 + 8
                                        = 308

So the sum of the values of 3 and 8 is 308.

Example 3:

    I am the smallest 2-digit number with 7 as my ones digit. What number am I?

Solution:

The smallest 2-digit number is 10

Ones digit = 7   ← given
Tens digit = 1

So the number is 17.

Name _____ Lesson 3.1

Write the answer.

1. What is the place value of 6 in 64?

   Answer: _____

2. What is the smallest 3-digit number?

   Answer: _____

3. In a 2-digit number, the tens digit is 5, and the ones digit is 2 more than the tens digit. What is the number?

   Answer: _____

4. What is the sum of the values of 3 and 9 in 389?

   Answer: _____

5. What is the smallest possible 3-digit number using the digits 4, 8, and 1?

   Answer: _____

6. I am the smallest 2-digit number with 8 as my ones digit. What number am I?

   Answer: _____

7. What is the largest possible 3-digit number using the digits 5, 6, and 2?

   Answer: _____

8. In a 2-digit number, the tens digit is 5, and the ones digit is 1 less than the tens digit. What is the number?

   Answer: _____

9. I am the largest 2-digit number with 4 as my ones digit. What number am I?

   Answer: _____

10. What is the place value of 5 in 58?

    Answer: _____

11. What is the sum of the values of 7 and 5 in 517?

    Answer: _____

12. What is the largest 3-digit number?

    Answer: _____

Name _____   ▶ Lesson 3.2

3.2 Place-Value Concepts – 2 (**)

Example 1:

What is the difference between the **place values** of 4 and 2 in 342?

Solution:

In the number 342:

    Place value of 4 = 10
    Place value of 2 = 1

Difference between place values of 4 and 2
                                     = 10 – 1
                                     = 9

So the difference between the place values of 4 and 2 is 9.

Example 2:

I am the smallest 3-digit number with 2 in the hundreds place. What number am I?

Solution:

We can find the answer by using a place-value table.

Hundreds	Tens	Ones

- The number has three digits. So there should be three boxes.

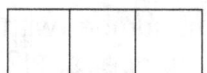

- The number is the smallest number, and 2 is the hundreds digit. This means we have 2 in the hundreds place. So write 2 in the hundreds place.

| 2 | | |

- The tens and ones place has to be filled with the smallest 1-digit number, which is 0.

| 2 | 0 | 0 |

So the number is 200.

Write the answer.

1. What is the sum of the place values of 6 and 8 in 681?

                          Answer: _____

2. What is the smallest possible 2-digit number using the digits 7 and 3?

                          Answer: _____

Name _____   Lesson 3.2

Write the answer.

3. What is the largest 2-digit number using the digits 3 and 8?

   Answer: _____

4. What number is 5 more than the value of the tens digit in 236?

   Answer: _____

5. I am the largest 3-digit number with 5 in the tens place. What number am I?

   Answer: _____

6. In a 3-digit number, the ones digit is 5. The tens and hundreds digits are 2 more than the value of the ones digit. What is the number?

   Answer: _____

7. What is the difference between the place values of 3 and 1 in 301?

   Answer: _____

8. What number is 3 less than the value of the hundreds digit in 567?

   Answer: _____

9. In a 2-digit number, the ones digit is 3. The tens digit is 5 more than the value of the ones digit. What is the number?

   Answer: _____

10. In a 3-digit number, the tens digit is 8. The ones and hundreds digit is 4 less than the value of the tens digit. What is the number?

    Answer: _____

11. What is the difference between the values of 9 and 4 in 794?

    Answer: _____

12. What is the smallest possible 3-digit number using the digits 9, 5, and 3?

    Answer: _____

3.3 Different forms to write number (*)

Example 1:

Write 586 in expanded form and find the missing number in the following math sentence.

586 = 500 + _____ + 6

Solution:

Write the numbers in the place value table.

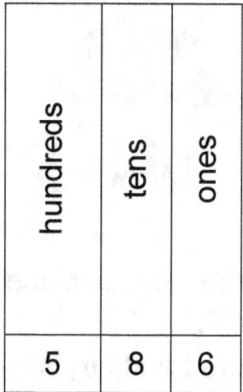

We can write the number 586 in expanded form is, 5 hundreds, 8 tens, and 6 ones.

5 hundreds = 500
8 tens = 80
6 ones = 6

586 = 500 + 80 + 6

So the missing number is 80.

Example 2:

I am given a number that can be written in expanded form as given below.

600 + 40 + 5

If I change 40 to 70, what will be the new number in standard form?

Solution:

First, we can write each number in the expanded form:

600 = 6 hundreds
40 = 4 tens
5 = 5 ones

Then write the numbers in the place value table.

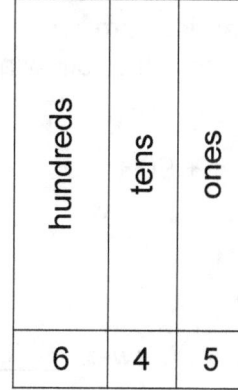

600 + 40 + 5 = 645

If I change 40 to 70 the new number will be,

600 + 70 + 5 = 675

So the new number in standard form will be 675.

Name _____   ▶ Lesson 3.3

Write or choose the letter of the answer.

1. Select the answer that shows the written form of 998.
   (a) Nine thousand ninety-eight
   (b) Nine hundred and ninety-eight
   (c) Nine hundred and ninety
   (d) None of the above

   Answer: _____

2. Write the expanded form given below in standard form:

   300 + 50 + 2

   Answer: _____

3. Write 798 in expanded form, and find the missing number in the following math sentence:

   798 = _____ + 90 + 8

   Answer: _____

4. I am given a number that can be written in expanded form, as given below:

   800 + 50 + 1

   If I change 800 to 200, what will be the new number in standard form?

   Answer: _____

5. Write the expanded form given below in standard form:

   80 + 2

   Answer: _____

6. Write 463 in expanded form, and find the missing number in the following math sentence:

   463 = 400 + _____ + 3

   Answer: _____

7. Select the answer that shows the written form of 857.
   (a) Eight hundred and fifty-seven
   (b) Eight thousand fifty-seven
   (c) Eight hundred and fifty
   (d) None of the above

   Answer: _____

8. I am given a number that can be written in expanded form, as given below:

   700 + 80 + 9

   If I change 700 to 600 and 9 to 4, what will be the new number in standard form?

   Answer: _____

Name _____   ▶ Lesson 3.4

3.4 Review of Chapter 3 – 1 (***)

Write or choose the letter of the answer.

1. In a 2-digit number, the tens digit is 7, and the ones digit is 3 less than the tens digit. What is the number?

   Answer: _____

2. I am the largest 2-digit number with 8 as my ones digit. What number am I?

   Answer: _____

3. Write the expanded form given below in standard form:

   600 + 50 + 2

   Answer: _____

4. What is the difference between the place values of 7 and 8 in 781?

   Answer: _____

5. What is the place value of 2 in 32?

   Answer: _____

6. What is the largest 2-digit number?

   Answer: _____

7. Select the answer that shows the written form of 593.
   (a) Five hundred and three
   (b) Five thousand ninety-three
   (c) Five hundred and ninety-three
   (d) None of the above

   Answer: _____

8. Write 874 in expanded form, and find the missing number in the following math sentence:

   874 = 800 + _____ + 4

   Answer: _____

9. What number is 4 more than the value of the hundreds digit in 236?

   Answer: _____

10. What is the sum of the values of 6 and 3 in 673?

    Answer: _____

Name _____ Lesson 3.4

Write or choose the letter of the answer.

11. Write 291 in expanded form, and find the missing number in the following math sentence:

    291 = _____ + 90 + 1

    Answer: _____

12. I am given a number that can be written in expanded form, as given below:

    900 + 80 + 7

    If I change 80 to 20, what will be the new number in standard form?

    Answer: _____

13. What is the smallest possible 3-digit number using the digits 5, 1, and 7?

    Answer: _____

14. Write 370 in expanded form, and find the missing number in the following math sentence:

    370 = 300 + _____ + 0

    Answer: _____

15. What is the sum of the place values of 8 and 5 in 850?

    Answer: _____

16. What is the largest possible 3-digit number using the digits 6, 8, and 4?

    Answer: _____

17. I am the largest 2-digit number with 0 as my ones digit. What number am I?

    Answer: _____

18. Select the answer that shows the written form of 602.
    (a) Six thousand two
    (b) Six hundred and twenty
    (c) Six hundred and two
    (d) None of the above

    Answer: _____

19. I am the smallest 3-digit number with 7 as my tens digit. What number am I?

    Answer: _____

Name _____   ▶ Lesson 3.5

3.5 Review of Chapter 3 – 2 (***)

Write or choose the letter of the answer.

1. What is the smallest 3-digit number?

    Answer: _____

2. What is the difference between the place values of 9 and 4 in 694?

    Answer: _____

3. Write 765 in expanded form, and find the missing number in the following math sentence:

    765 = _____ + 60 + 5

    Answer: _____

4. I am the largest 3-digit number with 5 as my hundreds digit. What number am I?

    Answer: _____

5. What is the place value of 4 in 341?

    Answer: _____

6. Write the expanded form given below in standard form:

    200 + 10 + 9

    Answer: _____

7. I am given a number that can be written in expanded form, as given below:

    600 + 40 + 7

    If I change 40 to 10 and 7 to 9, what will be the new number in standard form?

    Answer: _____

8. In a 3-digit number, the hundreds digit is 4, the tens digit is 5, and the ones digit has a value of 2 less than the tens digit. What is the number?

    Answer: _____

9. What number is 8 more than the value of the hundreds digit in 963?

    Answer: _____

Name _____ Lesson 3.5

Write or choose the letter of the answer.

10. Select the answer that shows the written form of 350.
    (a) Three hundred and five
    (b) Three thousand fifty
    (c) Three hundred and fifty
    (d) None of the above

    Answer: _____

11. Write 370 in expanded form, and find the missing number in the following math sentence:

    370 = 300 + _____ + 0

    Answer: _____

12. What is the sum of the place values of 7 and 9 in 709?

    Answer: _____

13. In a 3-digit number, the hundreds digit is 7, the ones digit is 1, and the tens digit has a value of 4 more than the ones digit. What is the number?

    Answer: _____

14. What is the largest 2-digit number?

    Answer: _____

15. Write the expanded form given below in standard form:

    200 + 10 + 9

    Answer: _____

16. I am given a number that can be written in expanded form, as given below:

    900 + 20 + 6

    If I change 900 to 300 and 20 to 90, what will be the new number in standard form?

    Answer: _____

17. I am the smallest 3-digit number with 8 as my tens digit. What number am I?

    Answer: _____

18. What is the largest possible 3-digit number using the digits 7, 8, and 5?

    Answer: _____

19. What is the sum of the values of 5 and 3 in 538?

    Answer: _____

Name _____   Lesson 4.1

4. Age Problems

4.1 Age Problems in the Present (*)

Example 1:

Bob is currently 15 years old. Kevin is 5 years older than Bob. How old is Kevin now?

Solution:

As given in the question:

Current age of Bob = 15 years

Current age of Kevin = 5 years older than Bob

We can find Kevin's current age as follows:

(Kevin's current age)
= (Bob's current age) + 5
= 15 + 5
= 20 years

So Kevin is 20 years old now.

Example 2:

The sum of Raul's and Vineet's ages is 28. If Raul is 12 years old, how old is Vineet?

Solution:

As given in the question:

Sum of Raul's and Vineet's ages = 28 years

Current age of Raul = 12 years

We can find Vineet's current age as follows:

Vineet's current age
= (sum of Raul's and Vineet's ages) − (Raul's current age)
= 28 − 12 = 16

So Vineet is 16 years old.

Write the answer.

1. The current age of Nancy is 18. If Maria is 11 years younger than Nancy, how old is Maria?

Answer: _____

2. Eli's current age is 11, and Mia's current age is 8. What is the sum of their current ages?

Answer: _____

Name _____

Lesson 4.1

Write the answer.

3. The current age of Pamela is 25. If Maya is 7 years younger than Pamela, how old is Maya?

    Answer: _____

4. David is currently 17 years old. Luke is 11 years older than David. How old is Luke now?

    Answer: _____

5. The sum of Ria's and Adriana's ages is 36. If Ria is 20 years old, how old is Adriana?

    Answer: _____

6. Emily's current age is 9, and Angela's current age is 15. What is the sum of their current ages?

    Answer: _____

7. The current age of Adam is 22. If John is 6 years younger than Adam, how old is John?

    Answer: _____

8. The sum of Peter's and Disha's ages is 42. If Peter is 18 years old, how old is Disha?

    Answer: _____

9. Max's current age is 15, and William's current age is 17. What is the sum of their current ages?

    Answer: _____

10. Sarah is currently 18 years old. Nancy is 10 years older than Sarah. How old is Nancy now?

    Answer: _____

11. The current age of Georgia is 26. If Sarika is 8 years younger than Georgia, how old is Sarika?

    Answer: _____

12. The sum of Ana's and Lia's ages is 29. If Lia is 13 years old, how old is Ana?

    Answer: _____

4.2 Age Problems in the Future (*)

Example 1:

Alex will be 15 years old in 4 years. How old is he now?

Solution:

As given in the question:

Alex's age in 4 years = 15 years

We can find Alex's current age as follows:

(Alex's current age)
 = (Alex's age after 4 years) − 4
 = 15 − 4 = 11 years

So Alex is 11 years old now.

Example 2:

Nathan is currently 7 years old. John is 5 years older than Nathan. What will be John's age in 6 years?

Solution:

We can solve this problem as given below:

Current age of Nathan = 7 years
Current age of John = 5 years older than Nathan

- Find John's current age.

  (John's current age)
   = (Nathan's current age) + 5
   = 7 + 5 = 12

- Find John's age in 6 years.

  To find John's age in 6 years, add 6 to his current age.

  (John's age in 6 years)
   = (John's current age) + 6
   = 12 + 6 = 18

So John will be 18 in 6 years.

Write the answer.

1. Jasmine is 20 years old. How old will she be in 2 years?

   Answer: _____

2. Simon will be 32 years old in 8 years. How old is he now?

   Answer: _____

Name _____   ▶ Lesson 4.2

Write the answer.

3. David is currently 10 years old. How old will he be in 6 years?

   Answer: _____

4. Dev will be 21 years old in 6 years. How old is he now?

   Answer: _____

5. Elina is currently 12 years old. Jacob is 15 years older than Elina. What will be Jacob's age in 3 years?

   Answer: _____

6. Pamela is currently 16 years old. How old will she be in 7 years?

   Answer: _____

7. Brian will be 10 years old in 5 years. How old is he now?

   Answer: _____

8. Kavya is 17 years old. How old will she be in 9 years?

   Answer: _____

9. Anuj is currently 11 years old. Amit is 4 years younger than Anuj. What will be Amit's age in 5 years?

   Answer: _____

10. Caroline is 24 years old. How old will she be in 2 years?

    Answer: _____

11. Michael is currently 10 years old. Olivia is 7 years older than Michael. What will be Olivia's age after 8 years?

    Answer: _____

12. Grace will be 25 years old in 5 years. How old is she now?

    Answer: _____

4.3 Age Problems in the Past (*)

Example 1:

Nelson was 7 years old in 2015. How old was he in 2012?

Solution:

As given in the question:

Nelson's age in 2015 = 7 years

We can use the following steps to answer the question:

Difference between 2015 and 2012
= 2015 − 2012
= 3 years

To find Nelson's age in 2012, subtract 3 years from his age in 2015.

(Nelson's age in 2012)
= (Nelson's age in 2015) − 3
= 7 − 3
= 4 years

So Nelson was 4 years old in 2012.

Example 2:

Nikita will be 17 years old in 2 years. How old was she 5 years ago?

Solution:

As given in the question:

Nikita's age after 2 years = 17 years

To find Nikita's current age, subtract 2 years from her age after 2 years.

Current age of Nikita
= Nikita's age after 2 years − 2
= 17 − 2
= 15 years

To find Nikita's age 5 years ago, subtract 5 years from her current age.

Nikita's age 5 years ago
= Nikita's current age − 5
= 15 − 5
= 10 years

So Nikita was 10 years old 5 years ago.

Write the answer.

1. Pater is currently 25 years old. How old was he 6 years ago?

    Answer: _____

2. Sonia was 13 years old 7 years ago. How old is Sonia now?

    Answer: _____

3. George will be 12 years old in 4 years. How old was he 2 years ago?

    Answer: _____

4. Alka was 10 years old in 2008. How old was she in 2005?

    Answer: _____

Name _____   Lesson 4.3

Write the answer.

5. Vijay is currently 28 years old. How old was he 11 years ago?

    Answer: _____

6. Jessica will be 14 years old in 5 years. How old was she 6 years ago?

    Answer: _____

7. Jack was 10 years old 3 years ago. How old is Jack now?

    Answer: _____

8. Anna is currently 22 years old. How old was she 7 years ago?

    Answer: _____

9. Samuel was 15 years old in 2008. How old was he in 2002?

    Answer: _____

10. Bill was 20 years old 6 years ago. How old is Bill now?

    Answer: _____

11. Molly was 6 years old in 2000. How old was she in 2013?

    Answer: _____

12. Juhi is currently 21 years old. How old was she 10 years ago?

    Answer: _____

13. Stella was 17 years old 2 years ago. How old is Stella now?

    Answer: _____

14. Frank will be 10 years old in 3 years. How old was he 4 years ago?

    Answer: _____

Name _____   ▶ Lesson 4.4

4.4 Review of Chapter 4 – 1 (*)

Write the answer.

1. Sonia will be 10 years old in 3 years. How old was she 2 years ago?

   Answer: _____

2. Bill is currently 12 years old. Kevin is 4 years older than Bill. What will be Kevin's age in 5 years?

   Answer: _____

3. Clara is currently 22 years old. How old was she 10 years ago?

   Answer: _____

4. Olivia is currently 18 years old. Molly is 7 years younger than Olivia. How old is Molly now?

   Answer: _____

5. Martin will be 25 years old in 8 years. How old is he now?

   Answer: _____

6. The sum of Amit's and Kunal's ages is 35. If Amit is 16 years old, how old is Kunal?

   Answer: _____

7. Juhi was 7 years old 6 years ago. How old is Juhi now?

   Answer: _____

8. The current age of Simon is 20. If Bob is 5 years younger than Simon, how old is Bob?

   Answer: _____

9. Lora is currently 17 years old. How old was she 3 years ago?

   Answer: _____

10. Linda will be 8 years old in 1 year. How old was she 3 years ago?

    Answer: _____

Name _____ ▶ Lesson 4.4

Write the answer.

11. Elina's current age is 16, and Mahi's current age is 12. What is the sum of their current ages?

    Answer: _____

12. Joseph is currently 10 years old. Alice is 7 years older than Joseph. How old is Alice now?

    Answer: _____

13. Anisha is 5 years old. How old will she be in 14 years?

    Answer: _____

14. Carlos is currently 15 years old. Steven is 9 years older than Carlos. What will be Steven's age in 10 years?

    Answer: _____

15. Helen will be 18 years old in 5 years. How old was she 6 years ago?

    Answer: _____

16. Alex is 11 years old. Nisa's age is the same as Alex's age. How old is Nisa?

    Answer: _____

17. Eric was 10 years old 5 years ago. How old is Eric now?

    Answer: _____

18. The sum of Angela's and Patricia's ages is 40. If Angela is 24 years old, how old is Patricia?

    Answer: _____

19. Maria will be 13 years old in 4 years. How old was she 6 years ago?

    Answer: _____

20. Birla will be 25 years old in 12 years. How old is he now?

    Answer: _____

4.5 Review of Chapter 4 – 2 (*)

Write the answer.

1. The current age of David is 22. If Charles is 4 years younger than David, how old is Charles?

    Answer: _____

2. Stella is currently 15 years old. John is 5 years older than Stella. What will be John's age in 6 years?

    Answer: _____

3. Martin is currently 7 years old. How old will he be in 8 years?

    Answer: _____

4. Julie's current age is 10, and Mia's current age is 16. What is the sum of their current ages?

    Answer: _____

5. Adriana was 13 years old in 2015. How old was she in 2009?

    Answer: _____

6. The sum of Brian's and Philip's ages is 38. If Brian is 18 years old, how old is Philip?

    Answer: _____

7. Grace will be 17 years old in 5 years. How old is she now?

    Answer: _____

8. Kavya is currently 14 years old. Jyoti is 9 years older than Kavya. How old is Jyoti now?

    Answer: _____

9. Kevin is currently 7 years old. How old will he be in 7 years?

    Answer: _____

10. Max will be 16 years old in 6 years. How old is he now?

    Answer: _____

Name _____

▶ **Lesson 4.5**

Write the answer.

11. Bimal is currently 25 years old. How old was he 6 years ago?

    Answer: _____

12. Angela will be 20 years old in 7 years. How old was she 3 years ago?

    Answer: _____

13. Nathan is currently 17 years old. How old will he be in 5 years?

    Answer: _____

14. Catherine was 12 years old in 2006. How old will she be in 2020?

    Answer: _____

15. Sofia is currently 13 years old. Salini is 3 years as old as Sofia. What will be Salini's age in 4 years?

    Answer: _____

16. The sum of Mark's and Luke's ages is 27. If Mark is 12 years old, how old is Luke?

    Answer: _____

17. Sarah's current age is 8, and Emily's current age is 13. What is the sum of their current ages?

    Answer: _____

18. Colin was 17 years old 2 years ago. How old is Colin now?

    Answer: _____

19. Thomas will be 18 years old in 4 years. How old is he now?

    Answer: _____

20. Christina is currently 24 years old. How old will she be in 6 years?

    Answer: _____

Name _____ ▶ Lesson 5.1

5. Travel Problems

5.1 Measuring Time (*)

Example 1:

It is 8 o'clock and dark outside. Is it nighttime or daytime?

(a) Nighttime
(b) Daytime

Solution:

It is 8 o'clock and dark.

When it is nighttime, it appears dark outside.

So the answer is nighttime, which is option (a).

Example 2:

It is 7 o'clock, and Serena is preparing dinner for her family. Is it a.m. or p.m.?

(a) A.M.
(b) P.M.

Solution:

Remember:

- Night 12 o'clock to noon 11:59 → a.m.
- Noon 12:01 to night 11:59 → p.m.

It is 7 o'clock now.

Dinner is usually prepared at night.

It is 7 o'clock at night, which is between noon 12:01 to night 11:59. So it is P.M.

The answer is P.M., which is (b).

Choose the letter of the answer.

1. It is 9 o'clock, and Mr. Henry is getting ready to go to his office. Is it a.m. or p.m.?

    (a) P.M.
    (b) A.M.

    Answer: _____

2. It is showing 8:45 on an analog clock, and people are coming back home from their offices. What time is it?

    (a) 8:45 a.m.
    (b) 8:45 p.m.

    Answer: _____

Name _____   ▶ Lesson 5.1

Choose the letter of the answer.

3. It is 11 o'clock, and children are going to sleep. Is it nighttime or daytime?

    (a) Daytime
    (b) Nighttime

    Answer: _____

4. It is showing 7:25 on an analog clock, and Carl is thinking about taking a shower and going to school. What time is it?

    (a) 7:25 p.m.
    (b) 7:25 a.m.

    Answer: _____

5. It is 1 o'clock, and employees at a company are taking a lunch break. Is it a.m. or p.m.?

    (a) A.M.
    (b) P.M.

    Answer: _____

6. It is 5 o'clock, and birds are coming back to their nests. Is it nighttime or daytime?

    (a) Daytime
    (b) Nighttime

    Answer: _____

7. It is 5 o'clock, and Lawrence is getting ready to go for a morning walk. Is it a.m. or p.m.?

    (a) P.M.
    (b) A.M.

    Answer: _____

8. It is 8 o'clock, and Mr. Woods went outside for a drive when the stars were shining. Is it nighttime or daytime?

    (a) Daytime
    (b) Nighttime

    Answer: _____

9. It is 10 o'clock, and sunny outside. Is it nighttime or daytime?

    (a) Nighttime
    (b) Daytime

    Answer: _____

10. It is showing 9:18 on an analog clock, and streetlights are on. What time is it?

    (a) 9:18 a.m.
    (b) 9:18 p.m.

    Answer: _____

5.2 Problems on Elapsed Time – 1 (*)

Example 1:

Roger looked at the clock, and the time was 7:00 a.m. If his school day starts at 9:00 a.m., how much time is there for him to get ready for school?

Solution:

We have to find how many hours there are between 7:00 a.m. and 9:00 a.m.

Start at 7:00 a.m. and count the number of hours until we reach 9:00 a.m.

There are 2 hours between 7:00 a.m. and 9:00 a.m.

So Roger has 2 hours to get ready for school.

Example 2:

If it is Friday today, what day will it be the day after tomorrow?

Solution:

Write all the days in the week.

Days in the Week
Monday
Tuesday
Wednesday
Thursday
Friday
Saturday
Sunday

Today, it is Friday.
Tomorrow, it will be Saturday.
The day after tomorrow, it will be Sunday.

So the day after tomorrow will be Sunday.

Write the answer.

1. If today is Sunday, what day will it be tomorrow?

    Answer: _____

2. Nikhil looked at the clock, and the time was 3:00 p.m. If his class ends at 4:00 p.m., how much time is left before the class ends?

    Answer: ____ _____
    unit

3. Vineet went to Los Angeles by bus. He looked at the clock, and the time was 8:00 a.m. If he is to arrive at 11:00 a.m., how much longer will it be before he reaches Los Angeles?

    Answer: ____ _____
    unit

4. If it was Thursday yesterday, what day is it today?

    Answer: _____

55

Name _____					Lesson 5.2

Write the answer.

5. It is 6:00 p.m., and Bivab wants to go shopping. His mother asked him to come back after 2 hours. When will Bivab come back?

   (a) 8:00 p.m.
   (b) 4:00 p.m.

   Answer: _____

6. If it will be Wednesday tomorrow, what day is it today?

   Answer: _____

7. James looked at the clock, and the time was 4:00 p.m. If his coaching class starts at 6:00 p.m., how many hours are left before coaching class?

   Answer: ____ _____
                    unit

8. If it will be Saturday tomorrow, what day was it yesterday?

   Answer: _____

9. Alice looked at the clock, and the time was 8:00 a.m. If she is going to go outside at 9:00 a.m., how much time is there to get ready?

   Answer: ____ _____
                    unit

10. It is 5:00 p.m., and Nikki wants to go to a movie with her friends. Her father asked her to come back after 4 hours. When will Nikki come back?

    (a) 1:00 p.m.
    (b) 9:00 p.m.

    Answer: _____

11. If it was Tuesday yesterday, what day will it be tomorrow?

    Answer: _____

12. Adam looked at the clock, and the time was 10:00 p.m. If he is going for a ride at 11:00 p.m., how much time is there to get ready?

    Answer: ____ _____
                      unit

5.3 Problems on Elapsed Time – 2 (**)

Example 1:

A movie started at 1:00 p.m. and continued for 2 hours and 30 minutes. When did the movie end?

(a) 4:00 p.m.
(b) 2:30 p.m.
(c) 3:30 p.m.
(d) 3:00 p.m.

Solution:

The movie started at 1:00 p.m. and continued for 2 hours and 30 minutes.

Start at 1:00 p.m. and count the number of hours and minutes.

Count 2 hours and 30 minutes to reach 3:30 p.m.

The movie ended at 3:30 p.m. So the answer is (c).

Example 2:

Mr. Jones went on leave for 5 days. If his leave started on Monday, what day will he come back?

(a) Friday
(b) Saturday
(c) Sunday
(d) Thursday

Solution:

Write all the days in the week.

Days in the Week
Monday
Tuesday
Wednesday
Thursday
Friday
Saturday
Sunday

The leave ends 5 days after Monday.

Count 5 days from Monday to reach Saturday.

Mr. Jones will come back on Saturday. So the answer is (b).

Choose the letter of the answer.

1. Nathan went on some official work for 2 days. If his work started on Saturday, what day will he come back?

    (a) Sunday
    (b) Friday
    (c) Monday
    (d) Thursday

    Answer: _____

2. A sports competition started at 3:00 p.m. and continued for 3 hours. When did the competition end?

    (a) 6:00 p.m.
    (b) 12:00 p.m.
    (c) 3:00 p.m.
    (d) 6:30 p.m.

    Answer: _____

Name _____ ▶ Lesson 5.3

Choose the letter of the answer.

3. Maria went to a circus at 5:00 p.m. and returned at 7:00 p.m. How long was the circus?

   (a) 5 hours
   (b) 2 hours
   (c) 7 hours
   (d) 9 hours

   Answer: _____

4. A quiz competition started at 10:00 a.m. and continued for 1 hour and 30 minutes. When did the competition end?

   (a) 11:30 p.m.
   (b) 8:30 p.m.
   (c) 8:30 a.m.
   (d) 11:30 a.m.

   Answer: _____

5. Mr. Tucker went to a seminar for 4 days. If his seminar started on Thursday, what day will he come back?

   (a) Tuesday
   (b) Saturday
   (c) Monday
   (d) Sunday

   Answer: _____

6. Carolyn went on a tour for 3 days. If her tour started on Tuesday, what day will she come back?

   (a) Friday
   (b) Sunday
   (c) Saturday
   (d) Thursday

   Answer: _____

7. Mrs. Collins went shopping at 5:00 p.m. and returned at 8:00 p.m. How long did she shop?

   (a) 8 hours
   (b) 2 hours
   (c) 6 hours
   (d) 3 hours

   Answer: _____

8. A school exam started at 8:00 a.m. and continued for 2 hours. When did the exam end?

   (a) 6:00 a.m.
   (b) 8:00 a.m.
   (c) 10:00 a.m.
   (d) 2:00 a.m.

   Answer: _____

5.4 Travel Problems – 1 (*)

Example 1:

Carter drove 75 kilometers in 1 hour. John drove 70 kilometers in 1 hour. How much farther did Carter drive than John in 1 hour?

Solution:

The following information is given:

Distance Carter drove in 1 hour
= 75 kilometers

Distance John drove in 1 hour
= 70 kilometers

To find how much farther Carter drove than John in 1 hour:

= (Distance Carter drove in 1 hour)

  – (Distance John drove in 1 hour)

= 75 – 70

= 5

So Carter drove 5 kilometers farther than John in 1 hour.

Example 2:

A tiger can run a certain distance in 20 minutes. A horse takes 3 minutes longer than the tiger to run the same distance. How long will it take the horse to complete the distance?

Solution:

The following information is given:

Time a tiger takes to run a certain distance = 20 minutes

Time a horse takes to run the same distance = 3 minutes longer than the tiger

To find the time it takes a horse to run the distance, add 3 minutes to the time it takes a tiger to run it.

Time a horse takes

= time a tiger takes + 3

= 20 + 3

= 23

So it will take the horse 23 minutes to complete the distance.

Write the answer.

1. Mia can walk a certain distance in 40 minutes. It takes Shane 4 minutes less than Mia to walk the same distance. How long will it take Shane to walk the distance?

   Answer: _____ _____
   unit

2. Marcus drove 54 kilometers in 1 hour. Peter drove 50 kilometers in 1 hour. How much farther did Marcus drive than Peter in 1 hour?

   Answer: _____ _____
   unit

Name _____   Lesson 5.4

Write the answer.

3. Martin was the last person to run in a race. His friends took 12 minutes to run the race. If Martin completed his part of the race in 10 minutes, how much longer did his friends take to complete the race?

   Answer: ____ _____
                      unit

4. Emily traveled 26 miles in 1 hour. Paul traveled 28 miles in 1 hour. How much farther did Paul travel than Emily in 1 hour?

   Answer: ____ _____
                      unit

5. Jack biked a certain distance in 54 minutes. It took Bill 11 minutes less than Jack to bike the same distance. How long did it take Bill to bike the distance?

   Answer: ____ _____
                      unit

6. Victor covered 40 kilometers in 1 hour. Mike covered 30 kilometers in 1 hour. How much farther did Victor travel than Mike in 1 hour?

   Answer: ____ _____
                      unit

7. Charles can travel a certain distance in 55 seconds. It takes Luke 2 seconds longer than Charles to travel the same distance. How long will it take Luke to travel the distance?

   Answer: ____ _____
                      unit

8. Monica was the last person to arrive at a party. Her colleagues arrived 5 minutes sooner than she did at the party. If it took Monica 25 minutes to arrive, how long did it take her colleagues to arrive the party?

   Answer: ____ _____
                      unit

9. Kelly walked 55 meters in 1 minute. Kate walked 65 meters in 1 minute. How much less of a distance did Kelly walk than Kate in 1 minute?

   Answer: ____ _____
                      unit

10. Nicole can run a certain distance in 10 minutes. It takes Jane 1 minute longer than Nicole to run the same distance. How long will it take Jane to run the distance?

    Answer: ____ _____
                      unit

5.5 Travel Problems – 2 (**)

Example 1:

Alastair can run 10 kilometers in 1 hour. If Michael can run 2 kilometers less than Alastair in the same amount of time, how far can Michael run in 1 hour?

Solution:

The following information is given:

Distance Alastair can run in 1 hour
= 10 kilometers

Distance Michael can run in 1 hour
= 2 kilometers less than Alastair

To find the distance Michael can run, subtract 2 kilometers from the distance Alastair can run.

Distance Michael can run
= Distance Alastair can run – 2
= 10 – 2
= 8 kilometers

So Michael can run 8 kilometers in 1 hour.

Example 2:

It takes Mr. Freeman 17 minutes to walk up the stairs and reach the topmost floor of a building. If it takes him 4 minutes less to walk down, how much time will it take him to walk down from the top floor to the ground?

Solution:

The following information is given:
Time it takes to walk up = 17 minutes
Time it takes to walk down
= 4 minutes less than to walk up

To find the time it takes to walk down, subtract 4 minutes from the time it takes to walk up.

Time it takes to walk down
= Time it takes to climb up – 4
= 17 – 4
= 13 minutes

So it takes Mr. Freeman 13 minutes to walk down from the top floor to the ground.

Write the answer.

1. Alex can walk 5 kilometers in 1 hour. If Hazel can walk 1 kilometer less than Alex in the same time, how far can Hazel walk in 1 hour?

Answer: _____ _____
unit

2. It takes Mr. Turner 28 minutes to travel to his office by bus. If it takes him 6 minutes less to travel to his office by car, how much time does he need to travel to his office by car?

Answer: _____ _____
unit

Name _____ ▶ **Lesson 5.5**

Write the answer.

3. It takes Robert 4 hours to climb up a hill with his crew. If it takes them 1 hour less to climb down, how much time do they need to climb down the hill?

   Answer: ____ _____
   　　　　　　　　unit

4. Jamie can swim 4 meters in 1 second. If Sarah can swim 1 meter more than Jamie in the same amount of time, how much farther can Sarah swim in 1 second?

   Answer: ____ _____
   　　　　　　　　unit

5. It takes Mrs. Dixon 11 minutes to make her lunch. If it takes her 5 minutes longer to make her dinner, how much longer does she need to make dinner?

   Answer: ____ _____
   　　　　　　　　unit

6. A boy can travel 35 kilometers in 1 hour. If a man can travel 55 kilometers in 1 hour, how much farther can the man travel than the boy in 1 hour?

   Answer: ____ _____
   　　　　　　　　unit

7. Frank can travel 35 miles in 1 hour. If Noah can travel 38 miles in 1 hour, how much farther can Noah travel than Frank in 1 hour?

   Answer: ____ _____
   　　　　　　　　unit

8. It takes Mr. Holmes 25 minutes to take his morning walk. If it takes him 5 minutes less to take his evening walk, how long does it take him to take his evening walk?

   Answer: ____ _____
   　　　　　　　　unit

9. A fish can swim 40 kilometers in 1 hour. If a boat can travel 50 kilometers in 1 hour, how much farther can the boat travel than the fish in 1 hour?

   Answer: ____ _____
   　　　　　　　　unit

10. Alan can drive 30 miles in 1 hour. If Shaun can drive 3 miles more than Alan in the same time, how far can Shaun drive in 1 hour?

    Answer: ____ _____
    　　　　　　　　unit

Name _____ Lesson 5.6

Write or choose the letter of the answer.

1. It is showing 7:20 on an analog clock, and people are coming back from their evening walk. What time is it?

   (a) 7:20 p.m.
   (b) 7:20 a.m.

   Answer: _____

2. Justin went to a museum by car. He looked at the clock, and the time was 8:00 a.m. If he arrives at 9:00 a.m., how long did it take to reach the museum?

   Answer: ____ _____
   unit

3. It takes Logan 28 seconds to climb a tree. If it takes him 4 seconds less to climb down, how much time does he need to climb down from the tree?

   Answer: ____ _____
   unit

4. It is 5 o'clock, and Ann is getting ready for after-school classes. Is it a.m. or p.m.?

   (a) A.M.
   (b) P.M.

   Answer: _____

5. It takes Mrs. Murphy 40 minutes to prepare dinner. It takes Mrs. Cooper 15 minutes longer than Mrs. Murphy to prepare dinner. How long will it take Mrs. Cooper to prepare dinner?

   Answer: ____ _____
   unit

6. Mr. Bailey went on a holiday tour for 6 days. If his tour started on Wednesday, what day will he come back?

   (a) Tuesday
   (b) Sunday
   (c) Wednesday
   (d) Monday

   Answer: _____

7. An essay competition started at 5:00 p.m. and continued for 2 hours. When did the competition end?

   (a) 3:00 p.m.
   (b) 1:00 p.m.
   (c) 7:00 p.m.
   (d) 9:00 p.m.

   Answer: _____

8. If it was Saturday yesterday, what day will it be tomorrow?

   Answer: _____

Name _____   Lesson 5.6

Write or choose the letter of the answer.

9. Thomas can swim 80 meters in 1 minute. If Colin can swim 10 meters less than Thomas in the same time, how far can Thomas swim in 1 minute?

   Answer: ____ _____
                    unit

10. Nicole looked at the clock, and the time was 4:00 p.m. If her dance class ends at 6:00 p.m., how much longer will it be before dance class ends?

    Answer: ____ _____
                     unit

11. It takes Ruby 8 minutes to make her breakfast. If it takes her 6 minutes longer to make her lunch, how much time does she need to make lunch?

    Answer: ____ _____
                     unit

12. Diana went to her uncle's house. She looked at the clock, and the time was 7:00 p.m. If she arrived at 8:00 p.m., how long did it take to reach her uncle's house?

    Answer: ____ _____
                     unit

13. It takes Grade 2 students 5 minutes to solve a math problem. If it takes Grade 3 students 1 minute less to solve the math problem, how long does it take Grade 3 students to solve the math problem?

    Answer: ____ _____
                     unit

14. It is 8 o'clock, and streetlights are on. Is it nighttime or daytime?

    (a) Nighttime
    (b) Daytime

    Answer: _____

15. Olivia went to a theater show at 8:00 p.m. and returned at 9:00 p.m. How long was the show?

    (a) 2 hours
    (b) 9 hours
    (c) 8 hours
    (d) 1 hour

    Answer: _____

16. Patrick traveled 92 kilometers in 2 hours. Tina traveled 85 kilometers in 2 hours. How much distance did Patrick travel than Tina in 2 hours?

    Answer: ____ _____
                     unit

Name _____ Lesson 5.7

5.7 Review of Chapter 5 – 2 (**)

Write or choose the letter of the answer.

1. If it will be Thursday tomorrow, what day is it today?

   Answer: _____

2. Max went to a science fair at 9:30 a.m. and returned at 11:30 a.m. How long was the fair?

   (a) 2 hours
   (b) 30 minutes
   (c) 11 minutes
   (d) 30 hours

   Answer: _____

3. It takes Scott 7 minutes to run around a park. If it takes him 13 minutes longer to walk around the park, how long does he need to walk around the park?

   Answer: ____ _____
   unit

4. It is 10 o'clock, and the sun is out. Is it a.m. or p.m.?

   (a) A.M.
   (b) P.M.

   Answer: _____

5. Gabriel can drive 43 miles in 1 hour. If Suzie can drive 10 miles less than Gabriel in the same time, how far can Suzie drive in 1 hour?

   Answer: ____ _____
   unit

6. It is 11 o'clock, and a group of men are in a forest for a picnic. Is it a.m. or p.m.?

   (a) P.M.
   (b) A.M.

   Answer: _____

7. Nathan went to a festival for 5 days. If the festival started on Friday, what day will he come back?

   (a) Thursday
   (b) Tuesday
   (c) Wednesday
   (d) Saturday

   Answer: _____

8. It is 8 o'clock, and Linda came home from her office. Is it nighttime or daytime?

   (a) Nighttime
   (b) Daytime

   Answer: _____

Name _____   Lesson 5.7

Write or choose the letter of the answer.

9. It takes Ian 10 minutes to press a shirt. If it takes him 3 minutes less to press a pair of trousers, how long does he need to press a pair of trousers?

   Answer: _____ _____
                      unit

10. It is showing 3:40 on an analog clock, and John is enjoying a magic show at an event. What time is it?

    (a) 3:40 p.m.
    (b) 3:40 a.m.

    Answer: _____

11. Edwin went to a village for survey. He looked at the clock, and the time was 2:40 p.m. If he arrived at 4:40 p.m., how long did it take him to reach the village?

    Answer: _____ _____
                       unit

12. Mr. Hughes went to a ministerial meeting at 2:00 p.m. and returned at 4:00 p.m. How long was the meeting?

    (a) 5 hours
    (b) 2 hours
    (c) 6 hours
    (d) 4 hours

    Answer: _____

13. Jimmy walked 4 miles in 1 hour. Trent walked 3 miles in 1 hour. How much farther did Jimmy walk than Trent in 1 hour?

    Answer: _____ _____
                       unit

14. It took Mr. Jordan 1 hour to go shopping. If it takes him 2 hours longer to go shopping with his wife, how long does he need to go shopping with his wife?

    Answer: _____ _____
                       unit

15. A pigeon can fly a certain distance in 30 minutes. It takes a dog 2 minutes more than the pigeon to run the same distance. How long does it take the dog to run the distance?

    Answer: _____ _____
                       unit

16. Chris looked at the clock, and the time was 5:00 a.m. If he slept and woke up at 7:00 a.m., how long did Chris sleep?

    Answer: _____ _____
                       unit

Name _____   ▶ Lesson 6.1

6. Money Problems

6.1 Shopping Problems (*)

Example 1:

Mr. Robinson bought 6 books for $15.00 and 5 notebooks for $6.00. How much money did he spend in total?

Solution:

The following information is given:

  Cost of 6 books = $15.00
  Cost of 5 notebooks = $6.00
  Total money spent
    = (cost of 6 books)
    + (cost of 5 notebooks)
    = $15.00 + $6.00
    = $21.00

So Mr. Robinson spent $21.00 in total.

Example 2:

Simon had $85.00. He spent $63.00 at a mall. How much money did he have after he finished shopping?

Solution:

The following information is given:

  Amount of money Simon had = $85.00
  Amount of money Simon spent = $63.00

Amount of money left
    = (amount of money Simon had)
    − (amount of money Simon spent)
    = $85.00 − $63.00
    = $22.00

So Simon had $22.00 after he finished shopping.

Write the answer.

1. Dan bought 8 wallets for $44.00. He gave $50.00 to the cashier. How much money did the cashier return?

    Answer: _____

2. BJ bought 4 buckets for $32.00 and 3 mugs for $15.00. How much money did he spend in total?

    Answer: _____

3. Disha had $61.00. She spent $21.00 in a hotel for dinner. How much money did she have after paying for dinner?

    Answer: _____

4. Jay had $74.00. He spent $54.00 while shopping in a mall. How much money did he have after he finished shopping?

    Answer: _____

Name _____ Lesson 6.1

Write the answer.

5. Elina bought 5 bottles for $15.00 and gave $20.00 to the cashier. How much money will the cashier return?

   Answer: _____

6. Mr. Smith bought 9 paintings for $72.00 and gave $100.00 to the cashier. How much money will the cashier return?

   Answer: _____

7. Amit bought 3 watches for $30.00 and 3 wallets for $15.00. How much money did he spend in total?

   Answer: _____

8. David bought 10 books for $13.00 and gave $20.00 to the cashier. How much money will the cashier return?

   Answer: _____

9. Nikhil bought 6 caps for $30.00 and 3 mufflers for $21.00. How much money did he pay in total?

   Answer: _____

10. Pamela bought 10 backpacks for $90.00 and gave $100.00 to the cashier. How much money will the cashier return?

    Answer: _____

11. Maya bought 8 books for $14.00 and 3 notebooks for $7.00. How much money did she pay in total?

    Answer: _____

12. If 5 bags cost $25.00 and 7 wallets cost $14.00, how much money do we need to buy those items?

    Answer: _____

13. Nancy had $66.00. She spent $41.00 on personal expenses at a mall. How much money did she have after shopping?

    Answer: _____

14. Paul bought 6 novels for $42.00 and gave $50.00 to the cashier. How much money will the cashier return?

    Answer: _____

Name _____   ▶ Lesson 6.2

6.2 Expense Planning (**)

Example 1:

Lucy wants to buy a ring that costs $87.00. She has $99.00. How much money will she have left to buy other things after purchasing the ring?

Solution:

The following information is given:

   Cost of a ring = $87.00

   Amount of money Lucy has = $99.00

Amount of money left

   = (amount of money Lucy has)
       − (cost of a ring)

   = $99.00 − $87.00

   = $12.00

So Lucy will have $12.00 to buy other things after purchasing the ring.

Example 2:

A cake costs $12.00, a burger costs $7.00, and a package of cookies costs $5.00. How much money do we need to buy these three things?

Solution:

The following information is given:

   Cost of a cake = $12.00

   Cost of a burger = $7.00

   Cost of a package of cookies = $5.00

Total amount of money needed

   = (cost of a cake)
   + (cost of a burger)
   + (cost of a package of cookies)

   = $12.00 + $7.00 + $5.00

   = $24.00

So we need $24.00 to buy these three things.

Write the answer.

1. Manoj wants to buy a television that costs $390.00. He also needs to buy a laptop that costs $700.00. How much money does he need in total?

                Answer: _____

2. Lora wants to buy a ring that costs $65.00. She paid $80.00. How much money did she get back?

                Answer: _____

3. A pizza costs $14.00, a burger costs $8.00, and a sandwich costs $5.00. How much money do we need to buy these three things?

                Answer: _____

4. Jack wants to buy a cooler that costs $155.00. He also needs to buy a computer that costs $700.00. How much money does he need in total?

                Answer: _____

Name _____ **Lesson 6.2**

Write the answer.

5. Mr. Smith wants to buy a pair of sandals that costs $32.00. He gave the cashier $35.00. How much money will the cashier return?

   Answer: _____

6. Nancy wants to buy a sewing machine that costs $80.00. She paid $90.00. How much money did she get back?

   Answer: _____

7. A chair costs $31.00, and a table costs $43.00. How much money do we need to buy these two things?

   Answer: _____

8. A burger costs $6.00, a packet of chocolates costs $10.00, and a pizza costs $12.00. How much money do we need to buy these three things?

   Answer: _____

9. A cycle costs $75.00. Jack wants to buy it. He gave $97.00 to the cashier. How much money will the cashier return?

   Answer: _____

10. Prabodh wants to buy a computer that costs $90.00. He has $95.00. How much money will he have to buy other things after purchasing the computer?

    Answer: _____

11. A pair of sunglasses costs $37.00, and a jacket costs $26.00. How much money do we need to buy these two things?

    Answer: _____

12. A packet of sweets costs $9.00, a packet of cashew nuts costs $16.00, and an ice cream cone costs $5.00. What is the cost of all three items?

    Answer: _____

6.3 Investment Problems (***)

Example 1:

Andy deposited an amount of $88.00 in his savings account. After a few days, his father deposited $45.00 in his account. What is the total amount in his account?

Solution:

The following information is given:

Amount of money Andy deposited = $88.00

Amount of money his father deposited = $45.00

Total amount of money

= (amount of money Andy deposited) + (amount of money his father deposited)

= $88.00 + $45.00

= $133.00

So Andy has $133.00 in his account.

Example 2:

Mr. Anderson donates $290.00 to a charity and $275.00 to a hospital every year. How much money does he donate in a year?

Solution:

The following information is given:

Amount of money donated to a charity = $290.00

Amount of money donated to a hospital = $275.00

Total amount of money donated

= (amount of money donated to a charity) + (amount of money donated to a hospital)

= $290.00 + $275.00

= $565.00

So Mr. Anderson donates $565.00 in a year.

Write the answer.

1. Jyoti had $89.00 in her bank account. After few days, she withdrew $57.00. How much money did she have left in her account?

    Answer: _____

2. Rob had $250.00 in his bank account and earned $2.50 in interest. How much money does he have in his account?

    Answer: _____

3. Mr. Ray donates $330.00 to a church and $450.00 to a charity. How much money does he donate?

    Answer: _____

4. Anuj had $92.00 in his savings account. His father deposited $58.00 in his bank account a few days later. What is the total amount in his account?

    Answer: _____

Name _____ Lesson 6.3

Write the answer.

5. Dev deposited an amount of $750.00 in his savings account. After one year, he earned $7.50 in interest. How much money does he have in his account?

    Answer: _____

6. Raul deposited an amount of $150.00 in his savings account. After a few days, his mother deposited $50.00 in his account. What is the total amount in his account?

    Answer: _____

7. Mr. Hill donates $400.00 to the less fortunate and $500.00 to a hospital. How much money does he donate?

    Answer: _____

8. Nikhil had $660.00 in his bank account and earned $6.60 in interest. How much money does he have in his account?

    Answer: _____

9. Kyle deposited an amount of $220.00 in his savings account. After one year, he earned $2.20 in interest. How much money does he have in his account?

    Answer: _____

10. Mrs. Clarke deposited an amount of $270.00 in her savings account. After a few days, her husband deposited $300.00 in her account. What is the total amount in her account?

    Answer: _____

11. Angela had $725.00 in her bank account and earned $7.25 in interest. How much money does she have in her account?

    Answer: _____

12. Nikita donates $200 to a temple and $150.00 to the less fortunate. How much money does she donate?

    Answer: _____

Name _____ ▶ Lesson 6.4

6.4 Pricing Problems (***)

Example 1:

Bob bought 7 cricket balls and paid a total of $48.00. He also bought 3 bats and paid a total of $79.00. What is the total cost of all the balls and bats?

Solution:

The following information is given:
  Cost of 7 cricket balls = $48.00
  Cost of 3 cricket bats = $79.00
Total cost of all the balls and bats =
  (cost of 7 cricket balls) +
  (cost of 3 cricket bats)
  = $48.00 + $79.00
  = $127.00

So Bob paid a total of $127.00 for all the balls and bats.

Example 2:

The cost of 5 pizzas is $18.00. Kamal gave some money to the cashier. If the cashier returned $2.00, how much money did Kamal give to the cashier?

Solution:

The following information is given:
  Cost of 5 pizzas = $18.00
Kamal gave some money to the cashier.
Amount of money cashier returned = $2.00
Amount of money Kamal gave to cashier
  = $18.00 + $2.00
  = $20.00

So Kamal gave $20.00 to the cashier.

Write the answer.

1. One bag of rice costs $9.00, and one box of cashew nuts costs $7.00. How much money do we need to buy one bag of rice and one box of cashew nuts?

   Answer: _____

2. The cost of a gold ring is $78.00. Julie gave $80.00 to the cashier. How much money will the cashier return?

   Answer: _____

3. The cost of one bottle of orange juice is $3.50. What is the cost of two bottles of orange juice?

   Answer: _____

4. The cost of 1 kilogram of iron is $21.00. What is the cost of 2 kilograms of iron?

   Answer: _____

Name _____   ▶ **Lesson 6.4**

Write the answer.

5. The cost of 2 pizzas is $24.00, and the cost of 2 burgers is $15.00. What is the total cost of all the pizzas and burgers?

   Answer: _____

6. Samir bought 3 wallets that cost $27.00 and 3 belts that cost $21.00. What is the total cost of all the wallets and belts?

   Answer: _____

7. Nikita bought 5 lunch boxes and paid a total of $30.00. She also bought 6 sets of glasses and paid a total of $45.00. What is the total cost of the lunch boxes and sets of glasses?

   Answer: _____

8. The cost of 5 books is $30.00, and the cost of 6 pens is $18.00. What is the total cost of all the books and pens?

   Answer: _____

9. The cost of 4 burgers is $20.00, and the cost of 3 pizzas is $24.00. What is the total cost of all the burgers and pizzas?

   Answer: _____

10. The cost of 1 diamond necklace is $238.00. Angela paid $250.00 to the cashier. How much money will the cashier return?

    Answer: _____

11. The cost of 1 kilogram of sugar is $4.00. If David buys 2 kilograms of sugar, then how much money will he pay the cashier?

    Answer: _____

12. Albert bought 8 books and paid a total of $80.00. He also bought 5 notebooks and paid a total of $15.00. What is the total cost of all the books and notebooks?

    Answer: _____

Name _____   ▶ **Lesson 6.5**

6.5 Profit and Loss (***)

Example 1:

   Bill bought a mobile phone for $78.00 and sold it for $12.00 more than what it cost. How much was the selling price?

Solution:

The following information is given:

   Cost of the mobile phone = $78.00

   Selling price = $12.00 more than the cost of the mobile phone

We can answer this question as follows:

The selling price = $78.00 + $12.00

   = $90.00

So the selling price is $90.00.

Example 2:

   Kavya spent $87.00 to buy 7 books. She sold them for a total of $90.00. How much of a profit did she make?

Solution:

The following information is given:

   Cost of 7 books = $87.00

   Selling price = $90.00

We can answer this question as follows:

Kavya's profit = (selling price)

   − (cost of 7 books)

   = $90.00 − $87.00

   = $3.00

So Kavya made a profit of $3.00.

Write the answer.

1. Jay bought a bicycle for $90.00 and sold it for $95.00. How much of a profit did he make?

                         Answer: _____

2. Bill bought 7 shirts for $78.00 and sold them for $71.00. How much of a loss did he take?

3. Samir bought a vacuum cleaner for $175.00 and sold it for $185.00. How much of a profit did he make?

                         Answer: _____

4. Disha bought a bicycle for $82.00 and sold it for $19.00 more than it cost. How much was the selling price?

                         Answer: _____

                         Answer: _____

Name _____

Lesson 6.5

Write the answer.

5. Edward spent $54.00 to buy 9 kilograms of rice. He sold the rice for $12.00 more than the buying price. How much was the selling price?

   Answer: _____

6. Carol bought 6 pairs of jeans for $66.00 and sold them for $70.00. How much of a profit did she make?

   Answer: _____

7. Diane bought a television for $156.00 and sold it for $150.00. How much of a loss did she take?

   Answer: _____

8. Jacob bought 5 markers for $25.00 and sold them for $13.00. How much of a loss did he take?

   Answer: _____

9. Nitin bought a bike for $920.00 and sold it for $110.00 more than it cost. How much was the selling price?

   Answer: _____

10. Georgia spent $97.00 on a cooler and sold it for $87.00. How much of a loss did she take?

    Answer: _____

11. Patricia spent $75.00 on a dress. She sold the dress for $16.00 less than what it cost. How much was the selling price?

    Answer: _____

12. Vinod bought 6 novels for $72.00. After a few days, he wanted to sell the novels. He sold them for $60.00. How much of a loss did he take?

    Answer: _____

Name _____   ▶ Lesson 6.6

6.6 Review of Chapter 6 (**)

Write the answer.

1. Sophia had $92.00. She spent $42.00 in a mall on her personal expenses. How much money did she have left after shopping?

    Answer: _____

2. Joseph bought 8 pairs of jeans for $88.00 and sold them for $70.00. How much of a loss did he take?

    Answer: _____

3. A burger costs $8.00, a packet of chocolate costs $12.00, and a pizza costs $14.00. How much money do we need to buy these three things?

    Answer: _____

4. 2 gold rings cost $212.00. Lily gave the cashier $250.00. How much money will the cashier return?

    Answer: _____

5. Anthony bought 10 books for $65.00. After a few days, he wanted to sell those books. He sold them for $60.00. How much of a loss did he take?

    Answer: _____

6. Ryan deposited $260.00 in his savings account. After a few days, his father deposited $110.00 in his account. What was the total amount in his account?

    Answer: _____

7. John deposited $422.00 in his savings account. After one year, he earned $40.00 in interest. What was the total amount in his account?

    Answer: _____

8. Julia bought a refrigerator for $256.00 and sold it for $233.00. How much of a loss did she take?

    Answer: _____

Name _____ Lesson 6.6

Write the answer.

9. Lucy bought a bicycle for $85.00 and sold it for $90.00. How much of a profit did she make?

Answer: _____

10. Morgan bought 8 books and paid a total of $48.00. He also bought 6 notebooks and paid a total of $24.00. What was the total cost of all the books and notebooks?

Answer: _____

11. Kameron bought a vacuum cleaner for $185.00 and sold it for $190.00. How much of a profit did he make?

Answer: _____

12. Ryker deposited $660.00 in his savings account. After a few days, his elder brother deposited $120.00 in his account. What was the total amount that was deposited in his account?

Answer: _____

13. Alina spent $66.00 to buy 11 kilograms of rice. He sold the rice for $15.00 more than the buying price. How much was the selling price?

Answer: _____

14. Olivia bought 6 lunch boxes and paid a total of $36.00. She also bought 8 sets of glasses and paid a total of $72.00. What was the total cost of the lunch boxes and the sets of glasses?

Answer: _____

15. One bag of rice costs $11.00, and one box of cashew nuts costs $12.00. How much money do we need to buy one bag of rice and one box of cashew nuts?

Answer: _____

16. Dalton deposited $490.00 in his savings account. After a few days, his elder sister deposited $150.00 in his account. What was the total amount that was deposited in his account?

Answer: _____

7. Mixture Problems

7.1 Mixture Problems with Objects (*)

Example 1:

Nathan has 15 blue pens and 10 red pens. If he combines all the pens, what will be the total number of pens?

Solution:

The following information is given:

    Number of blue pens = 15
    Number of red pens = 10

You can find the answer as shown below:

Total pens
    = Number of blue pens
    + Number of red pens
    = 15 + 10 = 25 pens

So Nathan has a total of 25 pens.

Example 2:

A basket has 14 apples. If we add 12 more apples to the basket, how many apples will there be in the basket?

Solution:

The following information is given:

    A basket has 14 apples
    12 more apples are added to the basket

You can find the answer as shown below:

Total apples
    = basket with 14 apples
    + 12 more apples
    = 14 + 12 = 26 apples

So there will be 26 apples in the basket.

Write or choose the letter of the answer.

1. Bag 1 and Bag 2 have a certain number of candies. Select the answer that equals a total of 56 candies in both of the bags.

    (a) 24 + 22
    (b) 31 + 35
    (c) 22 + 14
    (d) 15 + 41

    Answer: _____

2. Bob has 12 green marbles and 9 blue marbles. If he combines all the marbles, what will be the total number of marbles?

    Answer: ____ _____
                         unit

Name _____   Lesson 7.1

Write or choose the letter of the answer.

3. A basket has 28 packets of chocolate. If we add 19 more packets of chocolate to the basket, how many packets of chocolate will there be in the basket?

   Answer: ____ _____
   unit

4. Disha has 10 red soaps and 4 white soaps. If she combines all of the soaps, what will be the total number of soaps?

   Answer: ____ _____
   unit

5. Bag A and Bag B have a certain number of oranges. Some possible choices are given below. Select the choice that equals a total of 87 oranges in both the bags.

   (a) 53 + 38
   (b) 52 + 34
   (c) 42 + 44
   (d) 53 + 34

   Answer: _____

6. A package contains 14 pencils. If we add 7 more pencils to the package, how many pencils will the package contain?

   Answer: ____ _____
   unit

7. A box has 29 crayons. If we add 12 more crayons to the box, how many crayons will there be in the box?

   Answer: ____ _____
   unit

8. Box 1 and Box 2 have a certain number of paper boats. Some possible choices are given below. Select the choice that equals a total of 27 paper boats in both the boxes.

   (a) 19 + 8
   (b) 30 + 5
   (c) 11 + 15
   (d) None of the above

   Answer: _____

9. Andy has 41 pink candies and 19 red candies. If he combines all of the candies, what will be the total number of candies?

   Answer: ____ _____
   unit

10. A package has 43 red balloons and 27 blue balloons. If we combine all of the balloons, what will be the total number of balloons in the package?

    Answer: ____ _____
    unit

7.2 Mixture Problem with solutions (*)

Example 1:

Jar 1 has 40 liters of kerosene and jar 2 has 55 liters of kerosene. If we mix the content in the jars, what is the total quantity of kerosene?

Solution:

The following information is given:

  Amount of kerosene in jar 1 = 40 litters

  Amount of kerosene in Jar 2 = 55 litters

You can find the answer as shown below:

Total amount of kerosene
 = Amount of kerosene in jar 1
  + Amount of kerosene in jar 2
 = 40 litters + 55 litters
 = 95 litters

So the total quantity of kerosene is 95 litters.

Example 2:

Tank A has 20 liters of water. Tank B has some liquid made with 20 liters of water and 10 liters of other chemical. If we mix the liquids in both the tanks, what will be the total quantity of water?

Solution:

The following information is given:

  Amount of water in tank A = 20 liters

  Amount of liquid in tank B = 20 liters of water, and 10 liters of other chemical

So Amount of water in tank B = 20 liters

  Amount of chemical in tank B = 10 liters

- Total amount of water in both tank
  = Amount of water in tank A
    + Amount of water in tank B
  = 20 liters + 20 liters
  = 40 liters

So the total quantity of water will be 40 litters.

Write the answer.

1. A 90-milliliter bottle has 20 milliliters of alcohol. Simon wants to fill the rest of the bottle with water. How much water does he need?

   Answer: _____ _____

                               unit

2. Glass 1 has 120 milliliters of milk. Glass 2 has 70 milliliters of water. If we mix the contents of both of the glasses, what will be total amount of the solution?

   Answer: _____ _____

                               unit

Name _____ Lesson 7.2

Write the answer.

3. Glass A has 250 milliliters of juice. Glass B has some liquid that consists of 200 milliliters of juice and 100 milliliters of oil. If we combine the contents of both of the glasses, what will be the total quantity of juice?

   Answer: ____ _____
   unit

4. A 50-milliliter bottle has 30 milliliters of shampoo. Arushi wants to fill the rest of the bottle with water. How much water does she need?

   Answer: ____ _____
   unit

5. Glass 1 has 70 milliliters of alcohol. Glass 2 has 90 milliliters of a soft drink. If we combine the contents of both of the glasses, what will be the total amount of the drink?

   Answer: ____ _____
   unit

6. A 500-milliliter bottle has 250 milliliters of mango shake. BJ wants to fill the rest of the bottle with milk. How much milk does he need?

   Answer: ____ _____
   unit

7. Jar 1 has 14 liters of milk, and Jar 2 has 26 liters of milk. If we combine the contents of both of the jars, what is the total quantity of milk?

   Answer: ____ _____
   unit

8. Tank 1 has 87 liters of kerosine. Tank 2 has 71 liters of kerosene. If we mix the content in both the tanks, what will be the quantity of kerosene?

   Answer: ____ _____
   unit

9. Bottle A has 75 ml of shampoo and 25 ml of alcohol. Bottle B has 20 ml more shampoo than Bottle A. What is the amount of shampoo in Bottle B?

   Answer: ____ _____
   unit

10. Can A has 29 liters of milk and can B has 27 liters of milk. If we mix the content in both the cans, what is the total quantity of milk?

    Answer: ____ _____
    unit

Name _____ ▶ **Lesson 7.3**

7.3 Review of Mixture Problems (**)

Write or choose the letter of the answer.

1. A jar has 71 red marbles. If we add 27 green marbles to the jar, how many marbles will there be in the jar?

   Answer: ____ _____
                   unit

2. Bucket A has 10 liters of milk. Bucket B contains a liquid made of 12 liters of milk and 8 liters of water. If we combine the liquids of both of the buckets, what will be the total quantity of milk?

   Answer: ____ _____
                   unit

3. Basket 1 has 37 apples. Basket 2 has 17 oranges. If we combine the fruit in both of the baskets, how many pieces of fruit will there be?

   Answer: ____ _____
                   unit

4. A 100-milliliter bottle has 50 milliliters of honey. Kapil wants to fill the rest of the bottle with water. How much water does he need?

   Answer: ____ _____
                   unit

5. Packet A and Packet B have a certain number of crayons. Select the answer that equals a total of 34 crayons in both the packets.

   (a) 11 + 12
   (b) 17 + 18
   (c) 15 + 19
   (d) 10 + 16

   Answer: _____

6. Bottle 1 has 500 milliliters of juice and 100 milliliters of alcohol. Bottle 2 has 150 milliliters more juice than Bottle 1 and 200 milliliters of alcohol. If we combine the contents of both of the bottles, what is the total amount of solution?

   Answer: ____ _____
                   unit

7. Can 1 has 8 liters of water. Can 2 has 5 liters of vegetable oil. If we combine the contents of both of the cans, what will be the total amount of solution?

   Answer: ____ _____
                   unit

Name _____ Lesson 7.3

Write or choose the letter of the answer.

8. Drum 1 has 38 liters of liquid color. Drum 2 has 19 liters of water. If we combine the contents of both of the drums, what will be the quantity of liquid color?

   Answer: ____ _____
   　　　　　　　　unit

9. Tank A has 18 liters of coconut oil. Tank B contains a liquid made of 27 liters of coconut oil and 10 liters of gas. If we combine the liquids in both of the tanks, what will be the total quantity of coconut oil?

   Answer: ____ _____
   　　　　　　　　unit

10. A box has 17 biscuit packets. If we take 6 biscuit packets from the box, how many biscuit packets will there be in the box?

    Answer: ____ _____
    　　　　　　　　unit

11. Jyoti has 7 toy cars and 9 toy airplanes. If she combines the toys, what will be the total number of toys?

    Answer: ____ _____
    　　　　　　　　unit

12. An aquarium has 19 white fish and 17 red fish. If we combine the fish, what will be the total number of fish in the aquarium?

    Answer: ____ _____
    　　　　　　　　unit

13. Drum 1 has 87 liters of diesel oil. Drum 2 has 93 liters of kerosene. If we combine the contents of both of the drums, what will be the total quantity of diesel oil?

    Answer: ____ _____
    　　　　　　　　unit

14. Aquarium A and Aquarium B have a certain number of fish. Select the answer that equals a total of 43 fish in both the aquariums.

    (a) 12 + 30
    (b) 23 + 21
    (c) 19 + 24
    (d) 20 + 22

    Answer: _____

15. An 80-milliliter bottle glass has 70 milliliters of milk. How much of the glass is empty?

    Answer: ____ _____
    　　　　　　　　unit

7.4 Filling or Emptying a Tank – 1 (*)

Example 1:

A tap can add 65 liters of water to a tank in 1 minute. How much water can the tap add in 2 minutes?

Solution:

The following information is given:

    65 liters of water = 1 minute

In 2 minutes, the tap adds =

    1 minute = 65 liters
    2 minutes = 65 liters + 65 liters
               = 130 liters

So in 2 minutes, the tap can add 130 liters of water.

Example 2:

Pipe 1 takes 14 minutes to fill a container. Pipe 2 takes 11 minutes to fill the same container. How much less time does Pipe 2 take to fill the container than Pipe 1?

Solution:

The following information is given:

Time Pipe 1 takes to fill the container

    = 14 minutes

Time Pipe 2 takes to fill the container

    = 11 minutes

Less time Pipe 2 takes = time Pipe 1 takes
                       – time Pipe 2 takes

    = 14 minutes – 11 minutes
    = 3 minutes

So Pipe 2 takes 3 minutes less than Pipe 1 to fill the container.

Write the answer.

1. Two pipes are used for filling a milk drum. Pipe 1 can add 42 liters of milk in a minute. Pipe 2 can add 37 liters more than Pipe 1 in one minute. How much milk can Pipe 2 add in a minute?

    Answer: _____ _____
                           unit

2. A pump can add 54 liters of water to a tank in 1 second. How much water can it add in 2 seconds?

    Answer: _____ _____
                           unit

Name _____ ▶ Lesson 7.4

Write the answer.

3. A pump can add 82 liters of water to a tank in 30 seconds. How much water can it add in 1 minute?

   Answer: ____ _____
   unit

4. Two pipes are used for filling a tank. Pipe 1 can add 240 liters of liquid in 1 minute. Pipe 2 can add 45 liters more than Pipe 1 in 1 minute. How much liquid can Pipe 2 add in 1 minute?

   Answer: ____ _____
   unit

5. Pipe 1 takes 24 minutes to fill a water tank. Pipe 2 takes 33 minutes to fill the same water tank. How much more time does Pipe 2 take to fill the water tank than Pipe 1?

   Answer: ____ _____
   unit

6. A pump can empty Water Tank A in 54 minutes. It takes the same pump 5 minutes less to empty Water Tank B. How long does it take to empty Water Tank B?

   Answer: ____ _____
   unit

7. Two pipes are used for filling a tank. Pipe 1 can add 312 liters of liquid in 2 minutes. Pipe 2 can add 14 liters more than Pipe 1 in the same amount of time. How much liquid can Pipe 2 add in 2 minutes?

   Answer: ____ _____
   unit

8. A pump can add 42 liters of oil to a tank in 2 minutes. How much oil can it add in 4 minutes?

   Answer: ____ _____
   unit

9. A pump can fill Water Tank A in 18 minutes. It takes the same pump 7 minutes longer to fill Water Tank B. How long does it take the pump to fill Water Tank B?

   Answer: ____ _____
   unit

10. Pipe 1 takes 19 minutes to fill a container. Pipe 2 takes 29 minutes to fill the same container. How much less time does Pipe 1 take to fill the container than Pipe 2?

    Answer: ____ _____
    unit

7.5 Filling or Emptying a Tank – 2 (*)

Example 1:

Pump A takes 14 hours to empty a swimming pool. Pump B takes 2 hours longer to empty the same swimming pool. How long does Pump B take to empty the swimming pool?

Solution:

The following information is given:

Time Pump A takes to empty a swimming pool = 14 hours

How much longer Pump B takes to empty the same swimming pool = 2 hours

Find how long Pump B takes to empty the pool.

= Time Pump A takes
+ Time Pump B takes
= 14 hours + 2 hours
= 16 hours

So it takes Pump B 16 hours to empty the swimming pool.

Example 2:

Drum 1 can be filled in 26 minutes using a pipe. It takes the same pipe 5 minutes less to fill Drum 2. How long does it take the pipe to fill Drum 2?

Solution:

The following information is given:

Time the pipe takes to fill Drum 1
= 26 minutes

How much less time the pipe takes to fill Drum 2 = 5 minutes

Find how long it takes the pipe to fill Drum 2.

= Time the pipe takes to fill Drum 1
− How much less time the pipe takes to fill Drum 2
= 26 minutes − 5 minutes
= 21 minutes

So it takes the pipe 21 minutes to fill Drum 2.

Write the answer.

1. Pipe 1 can add 784 liters of water to a drum in 25 minutes, and Pipe 2 can add 269 liters of water in the same amount of time. If both pipes are opened for 25 minutes, how much water can be added to the drum?

   Answer: _____ _____
   unit

2. Tap A takes 15 minutes to fill a small pool. Tap B takes 5 minutes less to fill the same pool. How long does Tap B take to fill the small pool?

   Answer: _____ _____
   unit

Name _____ Lesson 7.5

Write the answer.

3. Tap 1 can add 350 liters of water to a drum in 1 hour, and tap 2 can add 320 liters of water in the same amount of time. If both taps are opened for 1 hour, how much water can be added to the drum?

   Answer: ____ _____
   unit

4. Tank 1 can be filled in 19 minutes using a pipe. The same pipe takes 8 minutes longer to fill Tank 2. How long does it take the pipe to fill Tank 2?

   Answer: ____ _____
   unit

5. Pipe A takes 10 hours to fill a pond. Pipe B takes 3 hours longer to fill the same pond. How long does it take Pipe B to fill the pond?

   Answer: ____ _____
   unit

6. Pipe 1 can add 415 liters of water to a drum in 1 hour, and Pipe 2 can add 397 liters of water in the same amount of time. If both pipes are opened for 1 hour, how much water can be added to the drum?

   Answer: ____ _____
   unit

7. Pipe A takes 10 minutes to fill an aquarium. Pipe B takes 3 minutes longer to fill the same aquarium. How long does Pipe B take to fill the aquarium?

   Answer: ____ _____
   unit

8. Pipe 1 can add 260 liters of water to a drum in 20 minutes, and Pipe 2 can add 187 liters of water in the same amount of time. If both pipes are opened for 20 minutes, how much water can be added?

   Answer: ____ _____
   unit

9. Pump 1 takes 10 hours to empty a swimming pool. Pump 2 takes 2 hours less than Pump 1 to empty the same swimming pool. How long does it take Pump 2 to empty the pool?

   Answer: ____ _____
   unit

10. Drum 1 can be filled in 18 minutes using a pipe. The same pipe takes 7 minutes less to fill Drum 2. How long does it take the pipe to fill Drum 2?

    Answer: ____ _____
    unit

Name _____   ▶ Lesson 8.1

8. Patterns

8.1 Basics of Number Patterns (*)

Example 1:

How many wheels are there on 4 bicycles?

Number of Bicycles:	1	2	3	4
Number of Wheels:	2	4	6	—

Solution:

In the table above, the number of bicycles and the number of wheels are given.

Find the number of wheels on 4 bicycles.

As given in the table above:

Number of wheels on 1 bicycle = 2
Number of wheels on 2 bicycles = 4
Number of wheels on 3 bicycles = 6

Each bicycle has 2 wheels. So count by 2 to find the number of wheels on 4 bicycles.

$6 + 2 = 8$ wheels

So there are 8 wheels on 4 bicycles.

Example 2:

How many pictures are there on 1 page?

Number of Pages:	4	3	2	1
Number of Pictures:	12	9	6	—

Solution:

In the table above, the number of pages and number of pictures are given.

Find the number of pictures on 1 page.

As given in the table above:

Number of pictures on 4 pages = 12
Number of pictures on 3 pages = 9
Number of pictures on 2 pages = 6

Count down by 3 to find the number of pictures on 1 page.

$6 - 3 = 3$ pictures

So there are 3 pictures on 1 page.

Name _____ Lesson 8.1

Write or choose the letter of the answer.

1.

Number of Cars:	1	2	3
Number of Wheels:	4	8	__

Which of the following is correct for the number of wheels on 3 cars?

(a) 12
(b) 10
(c) 16

Answer: _____

2. How many legs are there on 2 cows?

Number of Cows:	5	4	3	2
Number of Legs:	20	16	12	__

Answer: _____

3. How much water is there in 4 bottles?

Number of Bottles:	1	2	3	4
Amount of Water:	1	2	3	__

Answer: _____

4. How many pens are there in 1 box?

Number of Boxes:	4	3	2	1
Number of Pens:	16	12	8	__

Answer: _____

5. How many wheels are there on 4 tricycles?

Number of Tricycles:	1	2	3	4
Number of Wheels:	3	6	9	__

Answer: _____

6.

Number of Hands:	1	2	3
Number of Fingers:	5	10	__

Which of the following is correct for the number of fingers on 3 hands?

(a) 14
(b) 20
(c) 15

Answer: _____

Name _____ Lesson 8.2

8.2 Number Patterns in a Sequence (*)

Example 1:

Jack wrote a number pattern, as shown below:

2, 4, 6, 8, ____

What will be the next number in this pattern?

Solution:

This problem can be solved as given below:

In the above pattern, the rule is "Count by 2." That means each number is 2 more than the previous number.

The number before the missing number is 8.

The missing number = 8 + 2 = 10
So the next number in this pattern will be 10.

Example 2:

Mark wrote a number pattern, as shown below:

1, 3, 6, 8, 11, ____, ____

What will be the next two numbers in this pattern?

Solution:

This problem can be solved using the following steps:

Step 1: Find the rules of this pattern.

The first is "Count by 2," and the second is "Count by 3." These two rules are repeated in the above pattern, such as:
1 + 2 = 3, 3 + 3 = 6, 6 + 2 = 8, 8 + 3 = 11

Step 2: Find the value of the missing numbers.

By using the rules, the first missing number will be 2 more than 11, and the second missing number will be 3 more than the result.

The first missing number = 11 + 2 = 13
The second missing number is = 13 + 3
                              = 16

So the next two numbers in this pattern will be 13 and 16.

Write the answer.

1. Nikita wrote a number pattern, as shown below:

   12, 9, 6, ____, ____

   What will be the next two numbers in this pattern?

   Answer: _____

2. Alex wrote a number pattern, as shown below:

   5, 10, 20, 35, ____

   What will be the next number in this pattern?

   Answer: _____

Name _____   **Lesson 8.2**

Write the answer.

3. Jasmine wrote a number pattern, as shown below:

    8, 6, 4, ____

    What will be the next number in this pattern?

    Answer: _____

4. Frank wrote a number pattern, as shown below:

    2, 3, 5, 6, 8, ____, ____

    What will be the next two numbers in this pattern?

    Answer: _____

5. Mia wrote a number pattern, as shown below:

    2, 4, 8, 14, ____

    What will be the next number in this pattern?

    Answer: _____

6. Jay wrote a number pattern, as shown below:

    1, 4, 7, ____, ____

    What will be the next two numbers in this pattern?

    Answer: _____

7. Sarika wrote a number pattern, as shown below:

    1, 1, 2, 2, 3, ____, ____

    What will be the next two numbers in this pattern?

    Answer: _____

8. Bill wrote a number pattern, as shown below:

    5, 10, 15, ____

    What will be the next number in this pattern?

    Answer: _____

9. Adriana wrote a number pattern, as shown below:

    18, 14, 10, ____, ____

    What will be the next two numbers in this pattern?

    Answer: _____

10. Max wrote a number pattern, as shown below:

    3, 6, 12, 21, ____

    What will be the next number in this pattern?

    Answer: _____

Name _____   ▶ Lesson 8.3

8.3 Continuing Patterns Using Letters (*)

Example 1:

What is the next group of letters in the following pattern?

    A   AB   ABC   ____

Solution:

This problem can be solved as given below:

In the above pattern, the rule is "The next letter of the alphabet is added to the right of the previous group."

The missing group = ABCD

So the next group of letters in this pattern is ABCD.

Example 2:

What are the next two groups of letters in the following pattern?

    E   EE   EEE   ____ ____

Solution:

This problem can be solved as given below:

In the above pattern, the rule is "The same letter of the alphabet is added to the right of the previous group."

The first missing group = EEEE

The second missing group = EEEEE

So the next two groups of letters in this pattern are EEEE and EEEEE.

Example 3:

What is the next group of letters in the following pattern?

    A   BA   CBA   ____

Solution:

This problem can be solved as given below:

In the above pattern, the rule is "The next letter of the alphabet is added to the left of the previous group."

The missing group = DCBA

So the next group of letters in this pattern is DCBA.

Example 4:

What are the next two groups of letters in the following pattern?

    AB   BC   CD   ____ ____

Solution:

This problem can be solved as given below:

In the above pattern, the rule is "The next letter of the alphabet is added to the right of the last letter of the previous group."

The first missing group = DE

The second missing group = EF

So the next two groups of letters in this pattern are DE and EF.

Name _____   ▶ Lesson 8.3

Write the answer.

1. What is the next group of letters in the following pattern?

   P   QP   RQP   ____

   Answer: _____

2. What is the next group of letters in the following pattern?

   N   NN   NNN   ____

   Answer: _____

3. What is the next group of letters in the following pattern?

   R   SR   TSR   ____

   Answer: _____

4. What are the next two groups of letters in the following pattern?

   DE   EF   FG   ____   ____

   Answer: _____

5. What are the next two groups of letters in the following pattern?

   K   KL   KLM   ____   ____

   Answer: _____

6. What are the next two groups of letters in the following pattern?

   Q   QQ   QQQ   ____   ____

   Answer: _____

7. What is the next group of letters in the following pattern?

   M   MN   MNO   ____

   Answer: _____

8. What are the next two groups of letters in the following pattern?

   MN   NO   OP   ____   ____

   Answer: _____

9. What is the next group of letters in the following pattern?

   D   ED   FED   ____

   Answer: _____

10. What is the next group of letters in the following pattern?

    ST   TU   UV   ____

    Answer: _____

Name _____   ▶ Lesson 8.6

8.4 Find a Pattern (*)

Example 1:

　Robot's pattern is 1, 4, 7, 10 ___, ___

　(a) The rule in this pattern is "Count by _____."

　(b) What are the next two numbers?

Solution:

Pattern = 1, 4, 7, 10 ___, ___

In the above pattern, the rule is "Count by 3." That means each number is 3 more than the previous number.

So the answer for question (a) is 3.

The number before the missing number is 10.

The first missing number will be
　　　　　　= 10 + 3 = 13
The second missing number will be
　　　　　　= 13 + 3 = 16

So the next two numbers in this pattern are 13 and 16.

Example 2:

　What are the next two groups of letters in the following pattern?

　　ABC　DEF　GHI　___　___

Solution:

In the above pattern, the rule is "The next three letters of the alphabet create the next group."

The first missing group = JKL

The second missing group = MNO

So the next two groups of letters in this pattern are JKL and MNO.

Example 3:

　What is the next group of items in the following pattern?

　　A1　B2　C3　D4　___

Solution:

In the above pattern, the rule is "The next letter of the alphabet and the next cardinal number create the next group."

The missing group = E5

So the next group in this pattern is E5.

95

Name _____ ▶ **Lesson 8.4**

Write the answer.

1. Nil's pattern is 2, 4, 6, 8 _____

    (a) The rule in this pattern is "Count by _____."

    (b) What is the next number?

    Answer: (a) ____ (b) ____

2. What are the next two groups of items in the following pattern?

    AB1   BC3   AB1   BC3   ___   ___

    Answer: _____

3. Look at the pattern: 0, 2, 4, 6, 7, 10.
    There is one number that is wrong in the pattern.

    (a) The rule in this pattern is "Count by _____."

    (b) The wrong number in the pattern is _____

    (c) The correct number should be _____

    Answer: (a) ____ (b) ____ (c) ____

4. What is the next group of letters in the following pattern?

    MN   OP   QR   ___

    Answer: _____

5. What are the missing groups in the following pattern?

    D4   E5   F6   ___   ___

    Answer: _____

6. What are the next two groups of letters in the following pattern?

    GHI   JKL   MNO   ___   ___

    Answer: _____

7. What are the missing items in the following pattern?

    X5   Y6   X5   Y6   ___   ___

    Answer: _____

8. Look at the pattern: 1, 4, 7, 11, 13.
    There is one number that is wrong in the pattern.

    (a) The rule in this pattern is "Count by _____."

    (b) The wrong number in the pattern is _____

    (c) The correct number should be _____

    Answer: (a) ____ (b) ____ (c) ____

Name _____   Lesson 8.5

8.5 Review of Chapter 8 – 1 (*)

Write the answer.

1. What are the next two items in the following pattern?

    C2   D3   E4   ____   ____

    Answer: _____

2. What is the next group of letters in the following pattern?

    E   EF   EFG   ____

    Answer: _____

3. Amit wrote a number pattern, as shown below:

    1, 4, 8, 11, 15, ____, ____

    What will be the next two numbers in this pattern?

    Answer: _____

4. How many letters are there in 1 box?

Number of Boxes:	4	3	2	1
Number of Letters:	8	6	4	__

    Answer: _____

5. What is the next group of letters in the following pattern?

    K   LK   MLK   ____

    Answer: _____

6. What is the next group of letters in the following pattern?

    UV   VW   WX   ____

    Answer: _____

7. Nancy wrote a number pattern, as shown below:

    1, 2, 3, 4, ____, ____

    What will be the next two numbers in this pattern?

    Answer: _____

8. What is the next group of letters in the following pattern?

    EFG   HIJ   KLM   ____

    Answer: _____

Name _____   ▶ **Lesson 8.5**

Write or choose the letter of the answer.

9. Samir wrote a number pattern, as shown below;

   2, 4, 8, 10, 14 ____

   What will be the next number in this pattern?

   Answer: _____

10. What are the missing items in the following pattern?

    A10   B11   C12   ____   ____

    Answer: _____

11. What are the next two groups of letters in the following pattern?

    AB   BC   CD   ____   ____

    Answer: _____

12. Simon wrote a number pattern, as shown below:

    10, 7, 4, ____

    What will be the next number in this pattern?

    Answer: _____

13.

Number of Rooms:	1	2	3	4
Number of People:	3	6	9	__

Which of the following is correct for the number of people in 4 rooms?
   (a) 10
   (b) 12
   (c) 15

Answer: _____

14. What are the next two groups of letters in the following pattern?

    K   KL   KLM   ____   ____

    Answer: _____

15. What is the next group of letters in the following pattern?

    R   RR   RRR   ____

    Answer: _____

16. What are the next two items in the following pattern?

    U1   V2   W3   ____   ____

    Answer: _____

Name _____   Lesson 8.6

8.6 Review of Chapter 8 – 2 (*)

Write the answer.

1. Daniel wrote a number pattern as shown below

    1, 2, 5, 6, 9, ____, ____

    What will be the next two numbers in this pattern?

    Answer: _____

2. What is the next group of letters in the following pattern?

    B   CB   DCB   ____

    Answer: _____

3. Adam wrote a number pattern as shown below

    3, 6, 12, 21, ____, ____

    What will be the next two numbers in this pattern?

    Answer: _____

4. What is the next group of letters in the following pattern?

    MN   NO   OP   ____

    Answer: _____

5. Look at the pattern 1, 2, 3, 4, 6.

    There is one number that is wrong in the pattern.

    (a) The rule in this pattern is "Skip count by _____."

    (b) The wrong number in the pattern = _____

    (c) The correct number should be = _____

    Answer: (a) ____ (b) ____ (c) ____

6. Mark wrote a number pattern as shown below

    2, 4, 6, 8, 10, ____, ____

    What will be the next two numbers in this pattern?

    Answer: _____

7. What is the next group of letters in the following pattern?

    k   kk   ____

    Answer: _____

8. What are the next two groups of letters in the following pattern?

    AA1   BB2   CC3   ____   ____

    Answer: _____

Name _____ Lesson 8.4

Write or choose the letter of the answer.

9. How many caps are there in 1 marker?

Number of Markers:	6	5	4	3
Number of Caps:	6	5	4	__

Answer: _____

10. Mike's pattern is 1, 5, 9, 13 ____

   (a) The rule in this pattern is "Count by _____."

   (b) What is the next number?

Answer: (a) _____ (b) _____

11. Sarika wrote a number pattern, as shown below:

   2, 2, 4, 4, 6, ____, ____

   What will be the next two numbers in this pattern?

Answer: _____

12. What are the next two items in the following pattern?

   H6   I7   J8   ____ ____

Answer: _____

13. What is the next group of letters in the following pattern?

   G   HG   IHG   ____

Answer: _____

14. What is the next group of letters in the following pattern?

   Q   QR   QRS   ____

Answer: _____

15.

Number of Bikes:	1	2	3	4
Number of Wheels:	2	4	6	__

Which of the following is correct for the number of wheels on 4 bikes?
   (a) 8
   (b) 10
   (c) 6

Answer: _____

16. What is the next group of letters in the following pattern?

   CD   EF   GH   ____

Answer: _____

100

Quiz

1. It is showing 7:15 on an analog clock when a shopkeeper opens her shop in the morning. What time is it?

    (a) 7:15 p.m.
    (b) 7:15 a.m.

    Answer: _____

2. In a 3-digit number, the ones digit is 1. The tens and hundreds digits are the same and are 7 more than the ones digit. What is the number?

    Answer: _____

3. James had $34.00. He spent $18.00 on some books and the rest on a shirt. How much money did he spend on the shirt?

    Answer: _____

4. The cost of a wallet is $14.00, and the cost of a hat is $11.00. What is the total cost of the wallet and the hat?

    Answer: _____

5. A liquid cleaner contains 15 milliliters of liquid detergent and some water. If the amount of water is two times the amount of detergent, how much water does the cleaner have?

    Answer: ____ _____
                         unit

6. Two taps are used to fill a tank. Tap 1 can add 120 liters of liquid in 1 hour. Tap 2 can add 20 liters less liquid than Tap 1 in 1 hour. How much liquid can Tap 2 add in an hour?

    Answer: ____ _____
                         unit

7. Amy wrote a number pattern, as shown below:

    3, 6, 9, 12, ____, ____

    What will be the next two numbers in this pattern?

    Answer: _____

8. Alex is currently 21 years old. Emily is 3 years older than Alex. What is Emily's age?

    Answer: _____

Name _____   ▶ Quiz

9. Pipe 1 takes 18 hours to fill a pool. Pipe 2 takes 1 hour less to fill the same pool. How long does it take Pipe 2 to fill the pool?

   Answer: ____ _____

   unit

10. Select the answer that shows the word form of 624.
    (a) Six hundred and four
    (b) Six hundred and twenty-four
    (c) Six hundred and forty-two
    (d) None of the above

    Answer: _____

11. The sum of Mary's and Kim's ages is 32. If Kim is 18 years old, how old is Mary?

    Answer: _____

12. What is the next group of letters in the following pattern?

    AC   CE   EG   ____

    Answer: _____

13. What is the sum of the place values of 4 and 7 in 417?

    Answer: _____

14. If the following question has enough information, find the answer. Otherwise, write "No answer" for the answer.

    Kyle bought 6 storybooks. How much money did he give to the cashier?

    Answer: _____

15. If October 22 is on Friday, what day was it on October 20?

    Answer: _____

16. Frank can walk a certain distance in 42 minutes. It takes John 4 minutes longer than Frank to walk the same distance. How long will it take John to walk the distance?

    Answer: ____ _____

    unit

17. What is the math sentence for the following expression?

    Total of 8 and 11 taken away from 27
    (a) 27 − (8 + 11)
    (b) 27 + (8 + 11)
    (c) (27 − 11) + 8
    (d) (8 + 11) − 27

    Answer: _____

www.ingramcontent.com/pod-product-compliance
Lightning Source LLC
Chambersburg PA
CBHW081156180526

45170CB00006B/2099